Urban Sustainability

The Urban Sustainability Book Series is a valuable resource for sustainability and urban-related education and research. It offers an inter-disciplinary platform covering all four areas of practice, policy, education, research, and their nexus. The publications in this series are related to critical areas of sustainability, urban studies, planning, and urban geography.

This book series aims to put together cutting-edge research findings linked to the overarching field of urban sustainability. The scope and nature of the topic are broad and interdisciplinary and bring together various associated disciplines from sustainable development, environmental sciences, urbanism, etc. With many advanced research findings in the field, there is a need to put together various discussions and contributions on specific sustainability fields, covering a good range of topics on sustainable development, sustainable urbanism, and urban sustainability. Despite the broad range of issues, we note the importance of practical and policy-oriented directions, extending the literature and directions and pathways towards achieving urban sustainability.

The series will appeal to urbanists, geographers, planners, engineers, architects, governmental authorities, policymakers, researchers of all levels, and to all of those interested in a wide-ranging overview of urban sustainability and its associated fields. The series includes monographs and edited volumes, covering a range of topics under the urban sustainability topic, which can also be used for teaching materials.

Jinliu Chen · Paola Pellegrini

Densification as a Design Tool for Sustainable Urban Regeneration

Towards Socio-spatially Integrated Residential Communities

Springer

Jinliu Chen
Suzhou City University
Suzhou, Jiangsu, China

Paola Pellegrini
American University of Beirut
Beirut, Lebanon

ISSN 2731-6483 ISSN 2731-6491 (electronic)
Urban Sustainability
ISBN 978-981-95-1652-0 ISBN 978-981-95-1653-7 (eBook)
https://doi.org/10.1007/978-981-95-1653-7

This Springer imprint is published by the registered company Springer Nature Singapore Pte Ltd.
The registered company address is: 152 Beach Road, #21-01/04 Gateway East, Singapore 189721, Singapore

Preface

This book investigates densification as a strategic design tool for sustainable urban regeneration, focusing on China's rapid urbanization and emphasizing social sustainability as a critical lens for evaluation. The book proposes densification not merely as a spatial tactic but as a social project—one that can reconcile the rapid pace of China's urbanization with the needs of its communities.

Drawing on empirical observations of post-1970s urban communities built after the opening up and reform, and China's 14th Five-Year Plan, which prioritizes a people-centered approach to urbanization, the book focuses on pre-2000 residential neighborhoods—many now facing decay—with a particular focus on the city of Suzhou, selected by the national government for pilot initiatives in urban regeneration.

The core argument of the book is that planning in China demands rethinking for the task of regeneration, which the country must face after 40 years of accelerated urbanization. Regeneration practice should shift from the city-wide focus usual of urban expansion plans to the neighborhood level and move beyond local government's top-down actions and design-driven technical control, experimenting with socially adaptive frameworks.

Starting from this hypothesis and drawing on extensive literature and case-based evidence, the book proposes a link between people-oriented regeneration and sustainable densification, not reduced to mere floor-area ratios and infrastructure quotas. Furthermore, the book proposes that a systematic framework promoting people-driven decision-making should underpin the regeneration process in the specific context of China.

To address the link between densification and regeneration, this book illustrates these parts:

(1) Theoretical Discussion of the concepts of regeneration and densification within the broader framework of China's evolving urbanization. The book argues that densification is an integral aspect of urbanization and should be context-specific and adaptive to local conditions.

(2) Empirical Case Studies in the city of Suzhou, which is in one of the most developed areas of China, the Yangtse River Delta. These studies assess the impact of densification on urban regeneration. They include: (a) an analysis of spatial vitality in a densified neighborhood, comparing its pre- and post-densification conditions; (b) a resident satisfaction study comparing two neighborhoods different in morphology but similar in FAR; and (c) some urban design pilot projects targeting regeneration.

The study concludes that densification, when integrated with regeneration, can not only intensify land use and facilitate the transformations, but also enhance the quality of the built environment, and achieve socio-spatial empowerment through participatory processes, community self-governance, and neighborhood interactions. In doing so, densification can be not only a multiplier of square meters or GDP points, but a driver of social sustainability in residential communities crucial for the success of regeneration projects and the quality of its micro-public realms.

Suzhou, China Jinliu Chen

The original version of the book has been revised. A correction to this book can be found at https://doi.org/10.1007/978-981-95-1653-7_13

Acknowledgements

I would like to express my heartfelt gratitude to everyone who supported me throughout the journey of writing this book.

First and foremost, I deeply thank my wife, Xuanyu Liu, my two lovely daughters, and our parents, for their love, patience, and unwavering support. Without their encouragement and help, this book would not have been possible.

I am sincerely grateful to my primary Ph.D. supervisor and co-author, Dr. Paola Pellegrini, for her invaluable guidance and encouragement. Her work has been essential in shaping my research and this book.

I would like to thank my colleagues and mentors, Dr. Bing Chen, Dr. Jianxiang Huang, Dr. Yunqing Xu, Prof. Joon Sik Kim, Dr. Qiantao Zhang, and Dr. Shih-yang Kao, for their insights and contributions to my work. My gratitude also extends to my dear friends, Dr. Geng Ma, Dr. Wenquan Gan, Dr. Kunlun Ren, Dr. Gao Du, Dr. Yueming Yang, Yang An, and Dr. Puyue Gong—our friendship is truly invaluable.

A special thanks to my co-authors over the years, particularly Haoqi Wang, Pengchen Li, Wenkang Tian, and Kexin Xu, for their collaboration on the published papers. Their contributions greatly facilitated the research process. Lastly, I am grateful to the staff at Suzhou City University, Xi'an Jiaotong-Liverpool University, and the University of Liverpool for their efficient support and management throughout this project.

July 2025 Jinliu Chen

I want to thank Jinliu, my exceptional friends, my great family, and UPD at XJTLU for accompanying me on the amazing Chinese journey.

Paola Pellegrini
Associate Professor
the School of Architecture and Design
American University of Beirut, Lebanon

Contents

About the Authors

Dr. Jinliu Chen is an academic affiliated with Suzhou City University, holding a Ph.D. from the University of Liverpool, UK. His research focuses on advancing knowledge in urban studies and promoting sustainable practices in urban environments. Chen has published extensively in esteemed journals, including "Cities", "Sustainable Cities and Society", "Heritage Science", and "Tourism Management". His innovative research approaches, such as using quantitative urban analyses for sustainable regeneration strategies, demonstrate his commitment to enhancing urban living conditions through research-driven insights. Through his work, he has contributed significantly to the discourse on sustainable urban practices and well-being.

Paola Pellegrini is an Associate Professor at the School of Architecture and Design of the American University of Beirut, Lebanon and holds a Ph.D. in Urbanism from the University of Venice, Italy. She has held prominent academic positions, including Associate Professor at the Xi'an Jiaotong-Liverpool University, Suzhou China, Visiting Professor at the Universidade Estadual Paulista, Faculdade de Arquitetura, Bauru Brazil, and Visiting Associate at the Harvard Graduate School of Design, Cambridge USA. Her recent research focuses on sustainability in urbanization and the role of densification in the Chinese urban context. She has extensively published in international journals and edited several books. She is a registered Architect in Italy and has always combined academic research and professional practice in Urban Planning and Design.

Acronyms

5D	Density, Diversity, Design, Destination Accessibility, Distance to Transit
AHP	Analytic Hierarchy Process
BE	Built Environment
BS	Block Size
CCCS	Complete Community Construction Standard
CFI	Comparative Fit Index
CMIN/DF	Relative Chi-Square
DA	Destination Accessibility
DF	Function Diversity
DT	Distance to Transit
EFA	Exploratory Factor Analysis
ER	Enclosure Rate
ETDZ	Economic and Technological Development Zone
FAR	Floor Area Ratio
FSI	Floor Space Index
FYP	Five Year Planning
GDP	Gross Domestic Product
GFA	Gross Floor Area
GFI	Goodness of Fit Index
GI	Greenness Index
GIS	Geographic Information System
GSI	Ground Space Index
LC	Link Connectivity
LD	Land Use Diversity
LE	Land Use Evenness
NFI	Normed Fix Index
NTSP	National Territory Spatial Planning
OECD	Organization for Economic Co-operation and Development
POI	Point of Interest
PSPNet	Pyramid Scene Parsing Network

PVI	Potential Vitality Index
QoL	Quality of Life
RMR	Root Mean square Residual
RMSEA	Root Mean Square Error of Approximation
SDGs	Sustainable Development Goals
sDNA	Spatial Design Network Analysis
SEM	Structural Equation Model
SPSS	Statistical Product Service Solutions
TLI	Tucker-Lewis Coefficient
TOD	Transit Oriented Development
USD	United States Dollar

List of Figures

List of Tables

Chapter 1
The Relevance of the Research

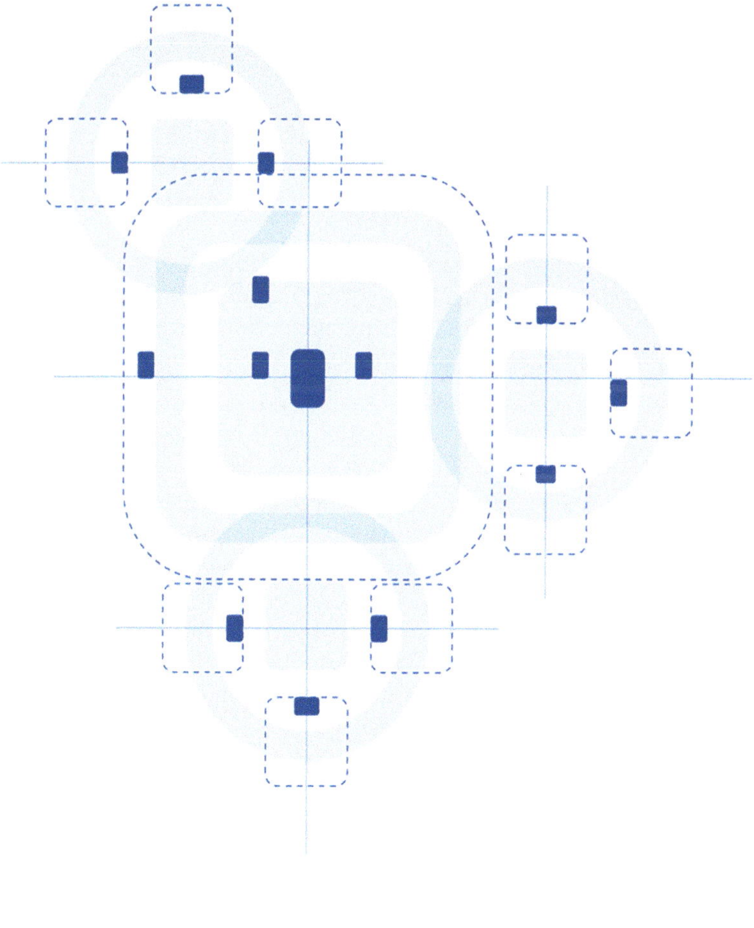

J. Chen and P. Pellegrini, *Densification as a Design Tool for Sustainable Urban Regeneration*, Urban Sustainability, https://doi.org/10.1007/978-981-95-1653-7_1

1.1 Research Topics

The study addresses three central concepts: sustainability, regeneration, and densification.

1. **Sustainability**: Sustainability is a huge issue, both for its relevance and for the wide spectrum of topics it is engaged with [1]. To be sustainable, in the multiple meanings this term has, is a worldwide requirement explicitly indicated by UN-Habitat [2]. The 2030 UN Agenda for Sustainable Development highlights 17 Sustainable Development Goals (SDGs) aimed at achieving global progress across social, environmental, and economic dimensions. In this research, sustainability is a pervasive subject.
2. **Regeneration**: With a big shift, recently China has decided to enter a regeneration phase in urbanization, at the same time re-considering what has already been built and planning what the future urban environment must be [3]. At the neighborhood level, which is the target of this book, regeneration entails improving indoor living conditions, increasing public facilities, and establishing a livable environment with comprehensive environmental, service, and governance systems.

   Specifically, the book targets the regeneration of the "old" residential neighborhoods—or communities—which are the ones built before 2000, because they are facing severe decay, and the 14th Five-Year Plan of Development of China promotes their regeneration.
3. **Densification**: Density is an essential parameter in the definition of every built environment. It is an evergreen issue every planner and designer must deal with, discussed over and over in disciplinary terms [4, 5]. As a concept bridging regeneration and sustainability, the advantages and disadvantages of various density levels and urban morphologies are subjects of discussion in the book.

This research investigates the meaning, goals, and strategies of densification, with a particular focus on the sustainable regeneration of "old" communities in China's recent urbanization. Through a combination of theoretical and empirical studies, this research proposes a densification framework aimed at enhancing spatial interventions and fostering people-oriented community regeneration. It proposes original research methods—evaluation, assessment, and verification tools—and applies these to several urban regeneration cases [6, 7].

The research methodology is mixed, aiming to capture the complexity of the built environment and its inhabitants. Beyond a critical review of existing literature, the methodology includes on-site observations, surveys, multi-source data collection with quantitative analysis, research-by-design exercises, and qualitative analysis to interpret the quantitative results [8, 9]. This mixing and the fact that the different research activities were not necessarily carried out one after the other in a consequential process, but were intertwined and overlapped, make a coherent narration of the research somehow difficult. The different parts of the book represent this complexity and research method [10].

1.2 Urban Regeneration in China as Research Background and Gap

In China, the concept of "high-quality" urbanization has shifted from a development-oriented model to a people-centered approach focused on sustainable regeneration. The 14th Five-Year Plan (FYP) (2021–2025) emphasizes urban regeneration (城市更新) as a key element in achieving sustainable and high-quality urbanization [3, 11]. This marks the first time that the central government has officially promoted urban regeneration as a strategy to optimize land use, enhance living conditions, and improve urban image. The 14th FYP defines urban regeneration as a broad concept, encompassing a range of activities—from demolition and reconstruction to partial and comprehensive renovation. Through these efforts, urban regeneration aims to adjust the urban configuration, aligning with the sustainability goals set by UN-Habitat in 2020 [12]. As China enters the regeneration phase of urbanization, the focus must shift towards achieving sustainable and high-quality growth [13, 14].

Research has shown that sustainable regeneration is shaped by a variety of factors, which differ based on individual needs and social environments. Therefore, it is not advisable to apply a one-size-fits-all approach to the regeneration of older residential communities [15–17]. Regeneration strategies should be dynamic, addressing the specific needs of residents within spatiotemporal contexts [18, 19]. Contemporary urban regeneration research has increasingly focused on governance mechanisms, financial support, and multi-party coordination [20–22]. Some studies have also emphasized improving residents' quality of life (QoL), particularly through community integration and strengthening neighborhood relationships [23].

From the outset of this research, it is clear that density influences both the physical form and residents' perceptions of their built environment [24]. Studies on the relationship between urban density and social sustainability have gained significant attention, particularly as old community regeneration is shaped by both physical conditions and residents' perceptions [25, 26]. Densification, which involves intensifying various forms and levels of density through multiple processes [27, 28], can result in more vibrant and active urban environments, integrating land use, housing, infrastructure, and culture [29]. Within the urban design discipline, but not only, recent research also highlights densification as an effective and flexible tool to optimize the spatial structure and hence achieve the spatial coherence to regenerate what exists [30, 31]. Moreover, urban assessment technologies and measurement methods play a crucial role in holistic regeneration decision-making, from evaluating the rationality of regeneration models to assessing residents' satisfaction and the vitality of regeneration outcomes [32]. Nevertheless, a systematic framework for densification remains underdeveloped, particularly in terms of its interdisciplinary application in the regeneration decision-making process [33].

At the micro-community scale in particular, regeneration is usually driven by project-based, result-oriented approaches controlled by local governments through spatial design [34]. In the absence of a comprehensive regeneration framework that integrates spatial intervention strategies to support sustainable, people-oriented decision-making for old community lies the research gap the study has explored [35]. Regenerating these communities requires attention to their unique identities and characteristics, rather than adopting a uniform approach [36]. Various stakeholders—researchers, officials, planners, and designers—tend to focus on their specific domains, such as policy-making, financial management, land use, and infrastructure planning [37] and regeneration decision-making lacks a unified approach that encompasses all these perspectives [38]. As people-centered urbanization continues to evolve and existing structures decay, old community regeneration demands a precise framework for sustainable upgrading.

1.3 Research Aim and Objectives

This research investigates densification as a tool for promoting projects of regeneration because the increase in density allows the transformation and the improvement of the qualities of the built environment. More specifically, with relevant literature reviews and empirical studies, the research explores sustainable densification for people-oriented regeneration of old residential neighborhoods. The conceptual and operational framework is identified based on residents' satisfaction evaluation and quantitative urban analysis. The research results can support the decision-making of local city planning authorities in their actions for sustainable urbanization.

To achieve the research aim, the research objectives are:

Objective 1: Analyze densification strategies for the sustainable regeneration of old communities, based on literature review and case studies.

Objective 2a: Critically assess densification by evaluating residents' satisfaction. This subjective evaluation measures, in a specific case study of regeneration, the outcomes of densification in a community.

Objective 2b: Critically identify key factors influencing residents' satisfaction with densification by conducting a comparative analysis of densification outputs in two different residential communities.

Objective 3: Perform a quantitative evaluation of densification cases using urban vitality assessments supported by the 5D theory, multi-source data, real-time photography technologies, and statistical tools such as the Analytic Hierarchy Process.

Objective 4: Synthesize insights from various case studies into a conceptual densification framework. Verify the framework drafting a pilot community regeneration project.

1.4 Research Questions

The essence of urban regeneration is to coordinate the relationship between space, environment, and people, and to build a livable, sustainable, intelligent, and resilient city. However, how to conduct old communities' regeneration is a research gap in China. Within this research gap, the role of densification in urban regeneration is yet to be explored. To fulfil the objectives, this research answers the following questions:

Research Question 1: Is densification a sustainable solution for old communities' regeneration in the Chinese urban environment?
Research Question 2: How can densification be successfully implemented under China's people-oriented urban regeneration goal at the community level?
Research Question 3: What comprehensive analytical system can assist densification decision-making and verify the outcomes for socially sustainable old community regeneration?
Research Question 4: What guidelines can be proposed for achieving sustainable densification at the community level?

1.5 Research Method Overview

The contemporary urban world is complex, and to address this complexity, mixed methodology is widely used in various disciplines [39, 40]. This complexity extends to community regeneration in China, where the decision-making, planning systems, and implementation processes demand an integrated approach. This study adopts a mixed research strategy of "evaluation, assessment, and verification" grounded in a comprehensive understanding of reality and aimed at interpreting people's attitudes [41]. The research applies a multi-faceted approach, combining qualitative and quantitative methods to explore community regeneration from diverse perspectives. It begins with a literature review and desktop research, followed by quantitative analysis to provide a well-rounded understanding. Ultimately, the study incorporates research by design, focusing on densification in an empirical context.

In the qualitative phase, the study synthesizes sustainable densification strategies through literature review and archival research. Drawing from China's new urbanization policies, urban regeneration movements, and direct observation, the research identifies gaps, objectives, and challenges of densification. It analyzes sustainability assessments within regeneration processes, incorporating current national planning frameworks and advanced architectural design methods in China. The conceptual framework is shaped by refining regeneration strategies at the community level. Field surveys and interviews are also conducted to understand residents' perceptions of their community environment. These insights inform the design of the questionnaire, focusing on residents' daily life, spatial preferences, and key environmental factors.

On the quantitative side, the study employs survey questionnaires and GIS-based urban analysis. The survey, analyzed through structural equation modelling (SEM), assesses residents' satisfaction in typical communities in Suzhou. This approach captures their feelings, opinions, and needs regarding community regeneration [42, 43]. Furthermore, a 5D model is used to assess urban vitality in Suzhou's regeneration projects [44]. Street view images, captured by wearable devices, quantitatively measure urban vitality, while the 5D model results serve as a verification tool. A highly feasible design strategy is generated through comprehensive and mixed research methods, and implementation densification suggestions and a framework for sustainable community regeneration are proposed [45, 46]. The study concludes by applying these methods in practical design cases, validating the framework and proposing strategies for implementation [47, 48].

1.6 Research Framework

To address the link between densification and regeneration, the study is structured in 12 chapters (Fig. 1.1), with in-depth theoretical analysis, followed by some empirical studies applied to real urban conditions, and concludes with an operational framework connected to a design proposal. The study has two main parts:

Part 1. The theoretical discussion on densification. The first part of the study presents and investigates the main topics of the research from a theoretical point of view: regeneration and densification, and their role in the general frame of increasing the sustainability of Chinese urbanization. After Chapter 1, which introduces the essential elements of the research, such as its aims, questions, and method, Chapter 2 introduces the idea of urban regeneration in China in the context of the "New Era" of sustainable development. Chapter 3 introduces the action of densification as a recurrent process in urbanization, explains in general terms its advantages and disadvantages, and highlights how density is context-related and should be context-driven. Chapter 4 explores the mutual relationship of the two research topics in increasing the sustainability of a specific kind of residential neighborhood: the communities where farmers were relocated when their villages were demolished to urbanize the land.

Part 2. The empirical case studies of densification. These studies investigate specific issues in specific places in Suzhou metropolitan area. The first one is the residents' satisfaction evaluation in Nanhuan: the residential community of Nanhuan was partially demolished, regenerated, and densified. The second one is a residential satisfaction evaluation comparison in Nanhuan and Daoqian communities to test the differential residents' perceptions of community regeneration [49]. The third case uses urban quantitative analysis via multi-source data and wearable equipment to measure Nanhuan community vitality [50], and investigates

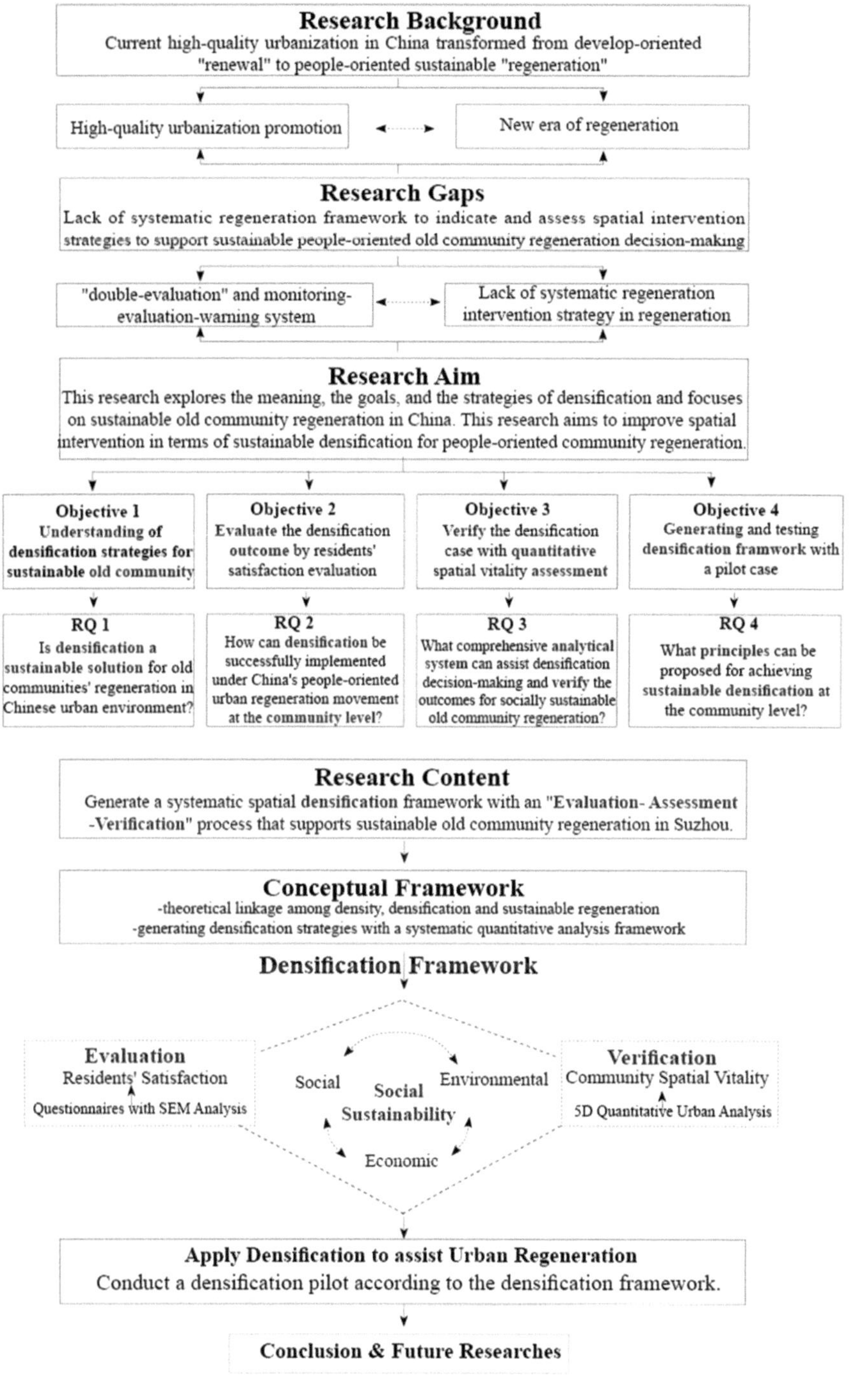

Fig. 1.1 Research framework

the underlying driving mechanism of urban vitality by combining the morphological and perceptual perspectives for shaping urban vitality efficiency in further regeneration decision-making [51].

1.7 Terminologies in the Chinese Context

Urban regeneration movement: This top-down national policy aims at land use optimization and preventing large-scale demolition and rebuilding [52]. It seeks to establish and refine sustainable regeneration patterns that promote long-term viability [53, 54].

"Old" community: The term refers to the residential neighborhoods built before the year 2000 in both cities and counties, impacted by the obsolescence of the structures, the poor maintenance, and the missing facilities, affecting the basic residents' Quality of Life (QoL) and their strong willingness to transform the neighborhoods [55].

Complete community: A well-rounded residential community offering comprehensive public service facilities, convenient commercial services, municipal infrastructure, ample public activity spaces, effective property management, and a robust community governance system. These elements ensure walkability and foster a strong sense of belonging and resident identity [59].

Double evaluation Mechanism: This mechanism assesses both the current state of land and space development and the effectiveness of spatial planning under the National Territory Spatial Planning framework. It includes pre-planning assessments and post-occupancy evaluations [56].

Densification: Densification involves the synergistic intensification of different forms and degrees of urban density. This process can enhance urban vibrancy, enabling mixed-use environments where infrastructure, public facilities, and open spaces are maximized [57]. In China, densification is integral to the urbanization process, concentrating populations in both existing and new urban areas [7, 58].

Regeneration: In China, urban regeneration focuses on revitalizing declining urban areas, replacing decaying structures with new functions and urban life [57]. It includes a wide range of actions, from demolition and redevelopment to small and targeted improvements.

References

1. Dale A, Onyx J (2010) A dynamic balance: social capital and sustainable community development. UBC Press, ISBN 0774859776
2. United Nations Department of Economic and Social Affairs (2017) The sustainable development goals report 2017 [R]. United Nations Publications, New York, pp 1–62. ISBN 978-92-1-101390-0. https://doi.org/10.18356/2671dabc-en. Accessed 30 June 2025

3. State Council of the People's Republic of China (2021) Outline of the 14th five-year plan (2021–2025) for national economic and social development and long-range objectives through the year 2035 [R]. Central Party Literature Publishing House, Beijing. Translated by CCPC Party History and Literature Research Institute (Official English version)
4. Zhao M, Liu N, Chen J, Wang D, Li P, Yang D, Zhou P (2024) Navigating post-COVID-19 social–spatial inequity: unravelling the nexus between community conditions, social perception, and spatial differentiation. Land 13(4):563
5. Hu J, Chen J, Li P, Yan J, Wang H (2024) Systematic review of socially sustainable and community regeneration: research traits, focal points, and future trajectories. Buildings 14:881. https://doi.org/10.3390/buildings14040881
6. Chen J, Gan W, Liu N, Li P, Wang H, Zhao X, Yang D (2024) Community quality evaluation for socially sustainable regeneration: a study using multi-sourced geospatial data and AI-based image semantic segmentation. ISPRS Int J Geo Inf 13(5):167. https://doi.org/10.3390/ijgi13050167
7. Chen J, Zhao X, Wang H, Yan J, Yang D, Xie K (2024) Portraying heritage corridor dynamics and cultivating conservation strategies based on environment spatial model: an integration of multi-source data and image semantic segmentation. Herit Sci 12:419. https://doi.org/10.1186/s40494-024-01497-7
8. Zhang X, Lu Y, Xu Y, Zhou C, Zou Y (2024) Governing regional inequality through regional cooperation? A case study of the Guangdong-Hong Kong-Macau greater Bay Area. Appl Geogr 162:103135. https://doi.org/10.1016/j.apgeog.2023.103135
9. Zhang X, Sun Y, Luan X (2023) Regulation flexibility and legitimacy building in governing intercity railways: the polymorphous role of the Chinese provincial government. J Urban Aff 1–16
10. Chen J, Li P, Wang H (2024) MSPA-based collaborative evaluation and planning governance of territory and natural environment in metropolitan agglomerations—a case study of EGIDDZ in Yangtze River Delta. J Resour Ecol 15:1263–1273. https://doi.org/10.5814/j.issn.1674-764x.2024.05.014
11. Ji M (2015) Urban housing security satisfaction in China: an HLM model and empirical study based on the 2012 Lian's public service quality of Chinese cities survey. E-Gov 150:67–75
12. United Nations Human Settlements Programme (2020) Unpacking the value of sustainable urbanization. UK-iLibrary
13. Chen J, Wang H, Yang Z, Li P, Ma G, Zhao X (2023) Comparative spatial vitality evaluation of traditional settlements based on SUF: taking Anren ancient town's urban design as an example. Sustainability 15:8178
14. Chen J, Pellegrini P, Yang Z, Wang H (2023) Strategies for sustainable urban renewal: community-scale GIS-based analysis for densification decision making. Sustainability 15:7901
15. Liu Z, Wang X, Ma J (2020) The impact mechanism of Beijing community public space on neighborhood communication in the transitional period: a comparative analysis of local residents and immigrants. Geogr Sci 40:69–78
16. Scott MJ, Lennon M, Douglas O (2019) Mainstreaming green infrastructure as a health-promoting asset. T Ctry Plan 88:151–156
17. Gu Z, Zhang X (2021) Framing social sustainability and justice claims in urban regeneration: a comparative analysis of two cases in Guangzhou. Land Use Policy 102:105224
18. Chen Y, Xiao L, Chen M, Jiang Y (2020) Preliminary exploration of approaches to community participatory planning: case study of 'New Qinghe Experiment' in Beijing. Urban Plan Forum 6
19. Liu J, Zhang Y, Ran A (2020) Research on the renewal and transformation of old districts in Beijing: based on characteristics-dilemma-policy analysis framework. Soc Gov Rev 10
20. Lin C, Du Y, Yue J, Wang J (2019) A research on the urban regeneration mechanism of multiple-stakeholder collaboration: a case study of Shenzhen. Urban Plan Forum 06:56–62
21. Huang L, Zheng W, Hong J, Liu Y, Liu G (2020) Paths and strategies for sustainable urban renewal at the neighbourhood level: a framework for decision-making. Sustain Cities Soc 55:102074

22. Zhu S, Li D, Jiang Y (2020) The impacts of relationships between critical barriers on sustainable old residential neighborhood renewal in China. Habitat Int 103:102232
23. Zhang S, Lin S, Li Z, Guo Y (2019) The impact of neighborhood environment perception on self rated health of Chinese urban residents: a case study of Wuhan City. Hum Geogr 34:38–46
24. Lau KK-L, Chung SC, Ren C (2019) Outdoor thermal comfort in different urban settings of sub-tropical high-density cities: an approach of adopting local climate zone (LCZ) classification. Build Environ 154:227–238
25. Dempsey N, Brown C, Bramley G (2012) The key to sustainable urban development in UK cities? The influence of density on social sustainability. Prog Plan 77:89–141
26. Dave S (2010) High urban densities in developing countries: A sustainable solution? Built Environ 36:9–27
27. Chen J, Li P, Lei Y, Zhang Y, Lai C, Chen B, Liu J, Schnabel MA (2025) Leveraging augmented reality for historic streetscape regeneration decisionmaking: a big and small data approach with social media and stakeholder participation integration. Cities. https://doi.org/10.1016/j.cities.2025.106214
28. Ma G, Paola P, Wu H, Han H, Wang D, Chen J (2023) Impact of land-use mixing on the vitality of urban parks: evidence from big data analysis in Suzhou, Yangtze River Delta Region, China. J Urban Plan Dev 149:4023045. https://doi.org/10.1061/JUPDDM.UPENG-4334
29. Marques B (2020) Improving community health and wellbeing through multi-functional green infrastructure in cities undergoing densification. Vic Univ Wellingt
30. Perez LP, Sales ALP, da Silveira JAR (2020) Climate change and the unsustainable urbanism in the municipality of João Pessoa, PB, Brazil. Sustain Debate 11:304–340
31. Long Y, Huang CC (2019) Does block size matter? The impact of urban design on economic vitality for Chinese cities. Environ Plan B Urban Anal City Sci 46:406–422
32. Yue W, Chen Y, Zhang Q, Liu Y (2019) Spatial explicit assessment of urban vitality using multi-source data: a case of Shanghai, China. Sustainability 11:638
33. Li Y, Zhang X, Xia C (2023) Towards a greening city: How does regional cooperation promote urban green space in the Guangdong-Hong Kong-Macau Greater Bay Area? Urban For Urban Green 86:128033
34. Hou L, Liu Y (2017) Life circle construction in china under the idea of collaborative governance: a comparative study of Beijing, Shanghai and Guangzhou. Geogr Rev Jpn Ser B 90:2–16
35. Chen J, Pellegrini P, Wang H (2022) Comparative residents' satisfaction evaluation for socially sustainable regeneration—the case of two high-density communities in Suzhou. Land 11:1483. https://doi.org/10.3390/land11091483
36. Shen T, Yao X, Wen F (2021) The urban regeneration engine model: an analytical framework and case study of the renewal of old communities. Land Use Policy 108:105571. https://doi.org/10.1016/j.landusepol.2021.105571
37. Ye L (2011) Urban regeneration in China: policy, development, and issues. Local Econ 26:337–347. https://doi.org/10.1177/0269094211409117
38. Zhou T, Zhou Y, Liu G (2017) Key variables for decision-making on urban renewal in china: a case study of Chongqing. Sustainability 9:370
39. Zhang X, Chen Y, Xu Y, Yue W, Tang C (2023) Does regional cooperation constrain urban sprawl? Evidence from the Guangdong-Hong Kong-Macao Greater Bay Area. Landsc Urban Plan 235:104742
40. Zhang X, Shen J, Sun Y, Zhou C, Yang Y (2023) Formation of city regions from bottom-up initiatives: investigating coalitional developmentalism in the Pearl River Delta. Ann Am Assoc Geogr 113:700–716
41. Jia Q, Zhessakov A (2021) Study on ecological evaluation of urban land based on GIS and RS technology. Arab J Geosci 14:1–8
42. Hao T, Zhao Q, Huang J (2023) Optimization of tree locations to reduce human heat stress in an urban park. Urban For Urban Green 86:128017
43. Huang J, Cui Y, Li L, Guo M, Ho HC, Lu Y, Webster C (2023) Re-examining Jane Jacobs' doctrine using new urban data in Hong Kong. Environ Plan B Urban Anal City Sci 50:76–93

44. Chen J, Li P, Wang H (2025) Assessment of influential factors on commute and life satisfaction in a historic campus-adjacent environment: evidence from a comparison study of twin cities. J Urban Plan Dev 151:4024065
45. Huang J, Cui Y, Chang H, Obracht-Prondzyńska H, Kamrowska-Zaluska D, Li L (2022) A city is not a tree: a multi-city study on street network and urban life. Landsc Urban Plan 226:104469
46. Hao T, Chang H, Liang S, Jones P, Chan PW, Li L, Huang J (2023) Heat and park attendance: evidence from "small data" and "big data" in Hong Kong. Build Environ 234:110123
47. Huang J, Jones P, He X (2022) Masks, ventilation and exposure time: a web-based calculator of indoor COVID 19 infection risk. Front Built Environ 8:986923
48. Huang J, Hao T, Liu X, Jones P, Ou C, Liang W, Liu F (2022) Airborne transmission of the delta variant of SARS-CoV-2 in an auditorium. Build Environ 219:109212
49. Zhang X, Li Y, Zhou C, Luan X, Yuan F (2023) Rescaling of the land regime in the making of city-regions: a case study of China's Pearl River Delta. Urban Stud 60:483–500
50. Zhang X, Zhou Z, Song Y (2024) Contested geopolitics of distribution in the city-region-building: promoting enclave industrial parks in the Pearl River Delta, China. Trans Inst Br Geogr 50(2):e12697
51. Yang N, Chen J, Ban L, Li P, Wang H (2024) Pre-planning and post-evaluation approaches to sustainable vernacular architectural practice: a research-by-design study to building renovation in Shangri-La's Shanpian House, China. Sustainability 16:9568
52. Pan L, Gan W, Chen J, Xian X (2024) Ecological planning for mitigating urban heat based on thermal network simulation: a case study of Guangzhou-Foshan metropolitan area. J Chin Urban For 22:10–19. https://doi.org/10.12169/zgcsly.2024.06.26.0004
53. Zhao X, Chen J, Li J, Wang H, Zhang X, Yu F (2025) Unraveling the renewal priority of urban heritage communities via macro-micro dimensional assessment—a case study of Nanjing City, China. Sustain Cities Soc 124:106317
54. Ni H, Li H, Li P, Yang J (2025) Exploring the spatiotemporal influence of community regeneration on urban vitality: unraveling spatial nonstationarity with difference-in-differences and nonlinear effect with gradient boosting decision tree regression. Sustainability 17
55. Song H, Chen J, Li P (2025) Decoding the cultural heritage tourism landscape and visitor crowding behavior from the multidimensional embodied perspective: insights from Chinese classical gardens. Tour Manag 110:105180. https://doi.org/10.1016/j.tourman.2025.105180
56. Gu H, Zhao Z (2025) Rethinking the Nexus between Urban Amenities and Migrants' Hukou Transfer Intentions in China: The Role of the Hukou Value. Eurasian Geogr. Econ. 66: 442–468, https://doi:10.1080/15387216.2024.2326447
57. Chen J, Ren K, Li P, Wang H, Zhou P (2024) Toward effective urban regeneration post-Covid-19: urban vitality assessment to evaluate people preferences and place settings integrating LBSNs and POI. Environ Dev Sustain. https://doi.org/10.1007/s10668-024-05378-8
58. Zhang X, Chen J, Wang H, Yang D (2025) From policy synergy to equitable greenspace: unveiling the multifaceted effects of regional cooperation upon urban greenspace exposure inequality in China's megacity-regions. Appl Geogr 174:103472. https://doi.org/10.1016/j.apgeog.2024.103472
59. Zhao X. Yu F, Zhang X, Chen J. Li P (2025) Assessing Urban Renewal Efficiency via Multi-Source Data and DID-Based Comparison between Historical Districts. npj Herit. Sci 13: 389. https://doi:10.1038/s40494-025-01964-9

Chapter 2
Urban Regeneration in China

J. Chen and P. Pellegrini, *Densification as a Design Tool for Sustainable Urban Regeneration*, Urban Sustainability, https://doi.org/10.1007/978-981-95-1653-7_2

2.1 Introduction

Rapid urbanization in China has driven extensive urban expansion, which has supported economic growth but also created numerous contradictions, including imbalanced spatial structures, insufficient public services, inadequate infrastructure, and poor living conditions [1]. In order to solve these contradictions, the 14th FYP of China (2021–2025) contains a series of policies for achieving a "new era" of sustainable and high-quality urbanization [2]. As an example, limiting the expansion of land for new construction in megacities, and promoting the Urban Regeneration Movement to enhance living quality in existing urban areas [3]. This research examines this evolving context, focusing on the following aspects.

- Identifying key issues in contemporary regeneration in China.
- Analyzing how the new national planning policy underpins high-quality, sustainable urban development.
- Exploring how national guidelines, such as the "Complete Residential Community Construction Guide" (January 2022), redefine regeneration principles [4].

2.2 The New Era of Chinese Urbanization

Over the past four decades, China's urbanization has driven massive population concentration in metropolitan areas (Fig. 2.1). By 2020, the urbanization rate reached 63.89%, with nearly 900 million people residing in cities [5]. Predictions from both the People's Bank of China and Morgan Stanley suggest that by 2050, urbanization will rise to 80% [6]. Simultaneously, the per capita housing area has expanded rapidly, growing from 5 m^2 to nearly 40 m^2 (Fig. 2.2) [7]. This increase in both population and housing space has fueled rapid urban building expansion [8, 9]. In response, the central government has imposed strict controls on land expansion, leading to the rise of ultra-high-density communities in urban areas [10].

As urbanization progresses, alongside a booming economy and rising expectations for living quality, the per capita housing area continues to grow substantially [11, 12] (Fig. 2.2). The government has complemented this growth with essential services, including education, commerce, and healthcare, to support improved living standards [13]. Consequently, the morphology of communities is evolving toward higher density, aligned with the increasing demands for residential and communal services [14].

Recently, urbanization has entered a new phase, with policies aimed at improving the poor quality of the existing built environment [16, 17]. From the 12th (2011–2015) to the 13th (2016–2020) national five-year plan, affordable housing projects in China focused on Shantytown Rebuilding (shantytowns are defined as informal, dilapidated, and lacking adequate infrastructure housing) [18]. Since 2008, more than 50 million shantytowns have been redeveloped in China [19, 20]. During the 13th five-year plan period, the target was 20 million units and from 2016 to August 2020, more than 23

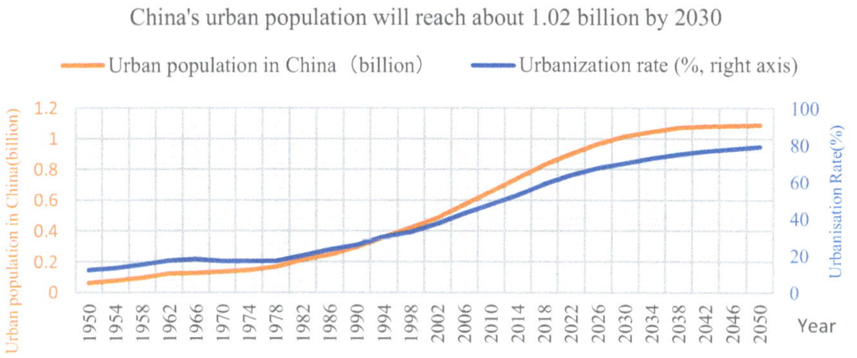

Fig. 2.1 The trend of urbanization rate of permanent residents in China; in 2019 and beyond are forecast values (*Source* Adapted form Morgan Stanley [15])

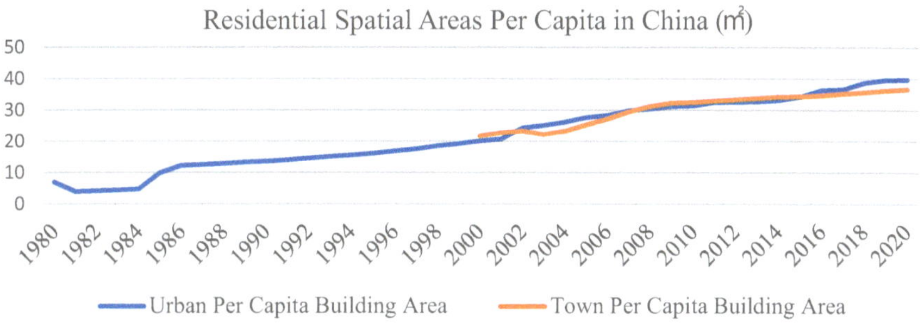

Fig. 2.2 Per capita residential area in cities and towns (*Source* Statistics adapted from the Ministry of Housing and Urban–Rural Development)

million shantytowns were renovated nationwide [21]. The acceleration and expansion of shantytown rebuilding are inseparable from the monetization policy of shantytown reform in 2013 (Fig. 2.3). The monetized resettlement of shantytown reform policy means that the government directly compensate the residents of the demolished shantytowns in cash [22] so that households buy houses in the commercial housing market.

The 14th five-year plan (2021–2025) did not mention the shantytown demolition and rebuilding policy [11, 20] but the urban regeneration movement to improve the living quality. Communities' built before 2000 have become the focus of the general office of the State Council guiding opinions on comprehensively promoting the transformation of old urban communities [23]. According to the 14th Five-year Plan and in combination with the actual conditions of various localities, China will strive to complete the transformation of old urban communities that need to be transformed [22, 24].

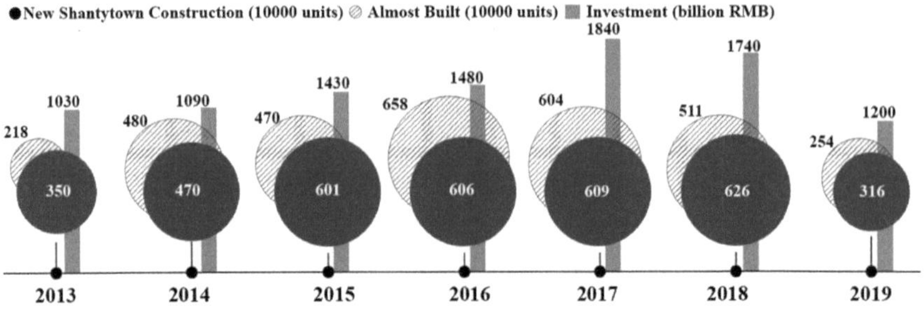

Fig. 2.3 The quantity and investment of various shantytowns in China in recent years (*Source* Adapted from the people's daily [14])

2.3 "Old" Community Features

In China, "old" community generally refers to the residential community built before 2000 for the obsolescence of its building quality and the lack of municipal facilities and public services [25, 26]. Major issues within the old communities include the following aspects.

2.3.1 *Large Base*

From 1978 to 2000, a total of 5.534 billion square meters of houses were built in China [27, 28]. The statistics of the construction volume before 1978 were not included because: (1) the limitation of information retrieval; (2) the service life standard of Chinese buildings was changed four times in 1966, 1974, 1989 and 2000, and most of the buildings built before 1980 exceeded the service life. At the same time, many low-quality old communities and dangerous dwellings were demolished.

At present, the information released by the Ministry of Housing and Urban Rural Development is that there were nearly 160,000 old residential communities in China before 2000, involving more than 42 million households and a construction area of about 4 billion square meters [29] (Fig. 2.4). This means a vast number of old community regeneration interventions is urgent [12, 30].

2.3.2 *Multiple Categories*

Old communities in the Chinese context can be summarized as government- and market-oriented. The government-oriented type could include the resettlement community [31], affordable housing [32], low-rent housing, housing cooperatives, etc. Market-oriented categories could include the urban village [33], purchased public housing, commercial housing [34], etc.

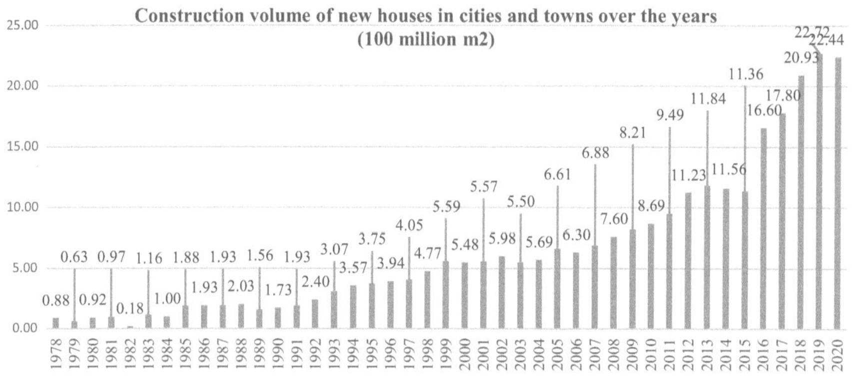

Fig. 2.4 Construction volume of new houses in cities and towns over the years (*Source* Adapted from Lianxun Securities [29])

2.3.3 Poor Environment

The problems in these old communities are prominent. First, a dilapidated pipe network for water supply and drainage, power grid, gas, and a lack of optical fiber facilities [19]. Additionally, common problems in the absence of public services include a lack of sufficient elderly and childcare facilities, inadequate parking, poor property management, a generally poor spatial environment, damaged roads. Last but not least, the low building quality, such as unstable structures, damaged external walls, infiltrated rainwater, and ageing roof insulation layers, is also a common problem in old communities [35]. The situation of the old community is complex, and there are three dilemmas to regenerating it:

- Governance Dilemma - Housing property rights are mixed and difficult to manage.
- System Dilemma - The amount of transformation required is large and all over China, but the construction systems and norms vary from place to place.
- Coordination Dilemma - Wide socio-economic differences, scarce and varied organizational approach, and struggle in getting consensus are relevant obstacles [36].

To address these dilemmas, the central and local governments in China recently promoted the urban regeneration movement, which is a series of actions to upgrade the old communities [37, 38].

2.4 Urban Regeneration Movement in China

The urban regeneration movement in China aims to renovate the declining areas in the city and meet the needs of the changing urban development. Renovation includes different approaches, also replacing the decaying physical space and introducing new urban functions to revitalize it, and promoting social equity [39]. The action has two targets: the objective transformation, such as the spatial, ecological, and visual environment; the continued transformation of soft elements such as cultural environments, recreational environments, neighborhood social networks, psychological settings, and emotional attachment [1].

Urban regeneration can be divided into three approaches: demolition and rebuild, rehabilitation and micro-renewal, and conservation [40]. Rehabilitation with micro-renewal and conservation are becoming significant parts of dealing with the old communities in China. However in the current laws and regulations system (Table 2.1), shantytown reconstruction and the so-called "three old reconstructions" are concepts assimilated to urban regeneration [41].

It must be highlighted that the regeneration of what exists, even if recently built, has been a continuous process in China since its opening up. The "making, remaking, and unmaking of urban landscape" method is essential to economic development [51]. So, in promoting "regeneration", the government is not promoting a new practice. What is innovative is the rethinking of urban qualities and resources that it expresses, the idea of building within the city limits, limiting the consumption of agricultural land, and producing a better urban environment than what was built in the last decades of the twentieth century.

In addition, the regeneration procedure is different than the new development on land yet to be urbanized, where the government decides what to do and leases the land to a real estate company for operation (Table 2.2). The urban regeneration development process is more complex for its multiple participants, the encouragement of public participation, and the need for private investors willing to invest in requalification [52].

2.5 New Goals in Urban Regeneration Decision-Making

2.5.1 People-Oriented High-Quality Urbanization: Sustainable Intensive Use of the Resources

During the rapid phase of urban expansion, demolition and rebuilding were implemented as quick and cost-effective methods for urbanization. However, as the urbanization rate nears 70%, the focus has shifted towards partial renewal and micro-transformations, rather than large demolition and redevelopment [53]. Protecting local communities has become a critical aspect of sustainable urban development, particularly within the framework of a circular economy [54]. Recently, the Ministry of Housing and Urban–Rural Development issued a document titled "Circular on Preventing Large-Scale Demolition and Construction in Urban Regeneration." This policy aims to curb the overdevelopment model that has led to new urban challenges, such as real estate speculation, rising housing prices, higher living costs, the destruction of historic urban structures, and the disruption of urban continuity.

Table 2.1 Main urban regeneration policies in China

Agency	Name	The core contents
Guangdong Government 2009.08	"Three old" reform policy	This policy clarifies how to operate for the regeneration of three categories of obsolete fabrics (cities and towns, factories, and villages); it includes land transfer and the improvement of the procedures. Previously such definitions and operations were not given [42]
Shenzhen Government 2009.12	Measures for urban renewal of Shenzhen	These measures can be considered the first systematic urban regeneration policy in China because they clearly define the principles, rules, and regulations of "government guidance, market operation, overall planning, economy and intensification, protection of rights and interests, and public participation". The policy establishes three renewal modes: Comprehensive Renovation, Functional Change, Demolition and Reconstruction [43]
Shenzhen Government 2010.09	Master Plan of Shenzhen City 2010–2020	This Master Plan is the first in China to adopt the spatial development mode from "incremental expansion (增量扩张)" to "optimization of what exists (存量优化)", emphasizing the shift from urbanization of new land to the regeneration of what is already built [44]
The national leadership 2013.12	The guidelines output of the conference on urbanization works	The guidelines establish that three major urban agglomerations in China—Pearl River Delta, Yangtze River Delta, and Beijing-Tianjin-Hebei—should focus on efficiently using existing resources and should no longer expand the construction land without restraint [45]
Ministry of land and resources 2014.02	Guidelines on strengthening control and implementing a stricter farmland protection system	The guidelines decide that, in principle, no additional construction land will be planned in the cities with a permanent urban population of more than 5 million (classified as megacities in China). These guidelines also define the "red line", a boundary of urbanization for avoiding new developments on fertile agricultural land [46]
The State Council 2014.03	National new urbanization plan (2014–2020)	The plan proposes a new type of urbanization for the whole Country, and it is the first programmatic guideline in China about high-quality urbanization, where new focuses are introduced, such as the "city for people" (以人为本) and "ecological civilization". The plan promotes green, circular, low-carbon development and requires avoiding urban sprawl, low efficiency in land use, the proliferation of useless urban districts, and over-sized wide roads [47]

(continued)

Table 2.1 (continued)

Agency	Name	The core contents
Shanghai Government 2014.03	Suggestions on further improving the level of economic and intensive use of land	The document puts forward some ideas for land management, including urban regeneration goals, such as "the maximum total amount of building quantity, the decrease of the growth rate of urbanization- incremental decline, stock optimization, capital flow efficiency and quality improvement."
The State Council 2016.02	Guidelines on strengthening the development of a new type of urbanization	The guidelines encourage the transformation of existing land resources through entrepreneurship and innovation and implement a policy of transforming land-use only according to its existing use and type of land rights (meaning land use can be changed into a new one only with a precise reason).
Shanghai Government 2016.02	Suggestions on further optimizing the structure of land and housing supply	The document encourages the renewal of existing commercial and office property and suggests increasing the proportion of new retail and office land that developers own in already existing development. This suggestion aims at inducing that offices and commercial construction are not built in large proportion and discouraging that commercial and office use is changed into residential. The proportion of self-owned office property and commercial property should not be less than 40% and 80%, respectively.
Ministry of Land and Resources 2016.04	Outline of the 13th Five-Year Plan for land and resources	The total amount of new land for construction in China will be kept at 32.56 million mu, as already decided in the previous Five-Year Plan. The area of land available for construction per unit of GDP will be reduced by 20% on average in every local Master Plan while allowing densification, increasing the proportion of existing construction surface for land supply [18]
The State Council 2016.06	Suggestions on accelerating the development of the housing rental market	The document allows to transform indoor spaces of buildings for residential leasing and allows commercial houses to be converted into rental houses in accordance with regulations [48]

(continued)

Table 2.1 (continued)

Agency	Name	The core contents
Shanghai Government 2017.11	Detailed rules for the implementation of urban renewal in Shanghai	The rules change the construction management concept to promote the city's organic regeneration. The urban action system of "1 + 4 + 12 + X" has been implemented; one urban design challenge, four action plans of shared community, innovation park, charm style, and leisure network; 12 confirmed pilot projects, and several (x) supplementary pilot projects around the action plan. It is required to strengthen the system and policy guarantee, and improve the dynamic management mechanism, and to expand the project participation platform and mobilize the strength of all sectors of society [49]
the CPC Central Committee 2019.12	Indications from the Central Committee, the output of the conference on economic development	It is necessary to guarantee the housing supply for low-income people, strengthen the urban regeneration and upgrading of existing housing, carefully transform the old residential areas in cities and towns, and vigorously develop rental housing. It is necessary to operate so that the house is used for living, not for speculation, and fully implement the long-term management and control mechanism of stabilizing the land price, house price, and expectation to promote the steady and healthy development of the real estate market [50]
the CPC Central Committee 2020.12	Indications from the Central Committee, the output of the conference on economic development	It is necessary to carry out the action of urban regeneration and promote the transformation of old residential areas in cities and towns. Some following related documents ([GBF [2020] No. 23]) define as "old" what was built before 2000 generally [23]

Source Government official webpages, collated by the authors

The idea is that high-quality urbanization is people-centered (Fig. 2.5), focusing on socially sustainable regeneration that respects local conditions and meet the residents' needs [55]. The idea is that through increased resident participation, self-governance, historical preservation, and equitable construction and management, sustainable urban regeneration can be achieved (Fig. 2.6).

Table 2.2 Urban regeneration procedure

Action in the process	Who does the action	Urban regeneration
Identify the potential transformation object	City master plan	The master plan identifies the areas of the "Three-olds" in the city: that is to say, the structures—old towns, old factories, and old villages—which are in poor condition because they do not perform according to the new standards of the number of people, of services, of environment.
Determine the purpose of the transformation	Local government	Multiple purposes are possible, related to the functions, the optimization of the building elements (façade, roof, accessibilities, facilities, etc.), the improvement of the environment, etc
Design the transformation	Design agency of the local government	A draft of the transformation details is produced that defines where and how the place must be transformed (regulatory plan), indicating land uses and density parameters.
Evaluate the implementation modes	Local government and real estate companies with inhabitants	In this phase, it is decided for in-situ resettlement of uprooted households or other solutions somewhere else together with the form of compensation for what is lost (in-kind compensation, transitional rent subsidies). The draft project is shown to people and published online to collect opinions and negotiation about solutions and compensations.
Calculate the compensation cost	A local government with the local inhabitants	The existing inhabitants must be compensated for losing their houses or for the temporary resettlement during the reconstruction. Different options of compensation are possible: the value of what was demolished in cash; an apartment of the existing residential surface or an increased one (1.5 or 2 times); renting compensation while waiting for the new unit (the new unit almost always has a higher real estate value than the original one).
Select investor	Local government Real estate company	The government agency in charge of the regeneration procedure searches for some private investors. The investor will bear the cost of resettlement compensation and the construction cost of resettlement housing. The real estate company will demand the local government's calculation of compensation and its related procedure. It can be the case that a private investor promotes the regeneration.
Calculate the transformation cost	Local government and Investment stakeholders	The transformation cost mainly includes the installation, demolition and construction price, temporary relocation cost, municipal fee, agricultural conversion cost, unforeseen cost, etc

(continued)

Table 2.2 (continued)

Action in the process	Who does the action	Urban regeneration
Select the implementation subject, builder	Local government	The project parts are assigned to a group of subjects or one according to the professional expertise (most of them are real estate enterprises) via open bidding procedure.
Check the process as planned	Investment stakeholders	Generally, the process is planned to be concluded after 5–8 years, and stakeholders actively work on this schedule.
Implementation assessment	Local government	After the transformation, some indicators, such as social stability risk assessment an' residents' satisfaction, are verified.

Source Urban regeneration projects in the cities mainly promoting urban regeneration, such as Shenzhen, Shanghai, Guangzhou, etc., synthesized by the authors

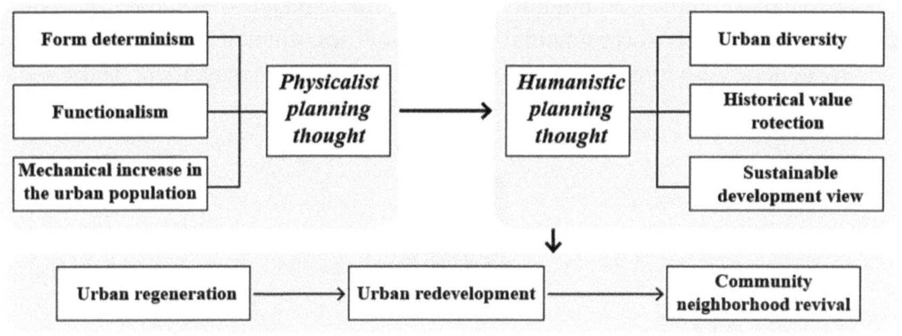

Fig. 2.5 The main notion on city planning shifted from physical planning to humanistic planning (*Source* Adapted by the authors from [56])

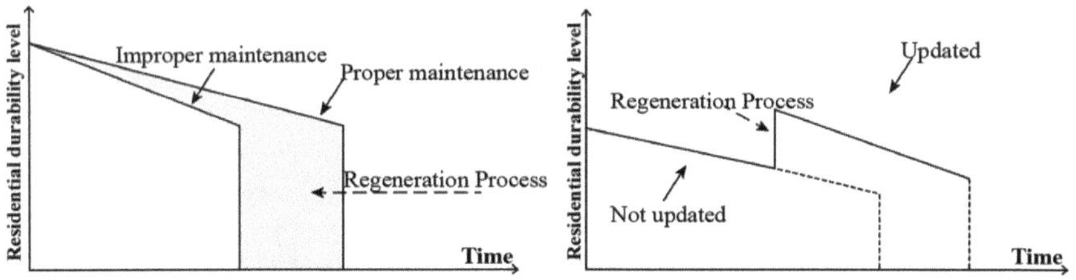

Fig. 2.6 The regeneration for upgrading the existing built environment quality (*Source* Image adapted by the authors from [56])

2.5.2 Complete Residential Community Guidelines: Improve Living Quality

The core of the urban regeneration initiative is people-oriented and multi-objective [57], encompassing: (1) urban function optimization, (2) environmental quality enhancement, and (3) property management improvement. Urban regeneration involves uncovering the potential for urban upgrading and redefining urban functions if needed. At its core, spatial regeneration aims to maximize land use efficiency. A key feature of this process is enhancing facility diversity [58].

In January 2022, the Complete Residential Community Construction Guide was issued by the Ministry of Housing and Urban–Rural Development (Fig. 2.7). Based on the investigation of over 200 old residential areas across 60 Chinese cities, the analysis of 7567 online questionnaires and the review of experiences from pilot cities like Xiamen and Shenyang the guide addresses critical challenges and outlines the future direction of residential community construction [59].

The goal of a complete community refers to sufficient and convenient basic public service facilities, such as commercial service facilities, municipal supporting infrastructure, general activity space, full coverage of industrial management, and a sound

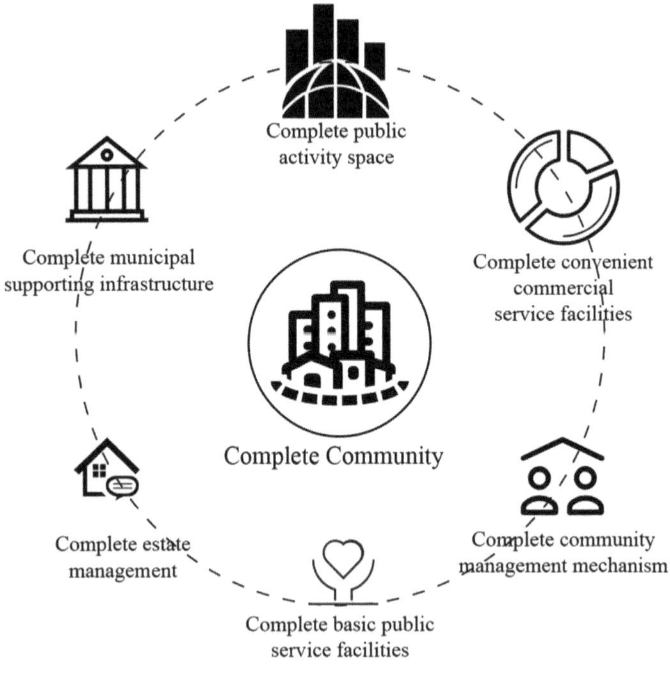

Fig. 2.7 Complete community contents (*Source* Image adapted by the authors from the Ministry of Housing and Urban–Rural Development [59])

community management mechanism, and a strong sense of belonging and identity (Table 2.3).

The urban regeneration movement is a sustainable urban regeneration policy which includes guidelines for the allocation of resources to realize the "complete community" [60], according to the requirements of the Standards for the Construction of Complete Residential Communities Guideline (Table 2.4). Through the construction of a complete residential community, a 15-min community-based life range is informed, enhancing the residents' livability.

Density plays an essential role in defining the "complete community," and therefore, it is necessary to understand and analyze the existing densities to guide the establishment of sustainable urban form. Taking a community of 50,000 people in 170,000 square meters as an example, to meet the requirements of a complete community means that the density of public service facilities needs to reach 3% [61].

If the community needs to meet the requirements of convenience commercial service facilities, the density of convenience commercial service facilities must reach 1%. If the community needs sufficient public space, the density of public space needs to reach 3% [62].

In summary, this insight of the complete residential community guide serves as a valuable reference for regenerating old residential areas. It adopts a problem-oriented approach, highlighting the shortcomings in the current construction of old residential communities. It also offers supportive facilities and public space standards aimed at improving both the built environment and residents' QoL.

Table 2.3 The content of complete residential community implementation

Implementation objectives	Implementation content
Complete basic public service facilities	Community service station, Kindergarten, Nursery, Elderly service station, Community health service station
Complete commercial service facilities	Comprehensive Supermarket, Multiple mails and express delivery facilities, and Other convenient commercial outlets
Complete municipal supporting infrastructure	Water, electricity, road, gas, heat, telecommunications and other facilities, Parking and charging facilities, Slow motion system, Wheelchair Accessible, Environmental sanitation facilities
Complete public space	Public activity space, public green space
Complete community management	Property services, property management service platform
Complete community management mechanism	Management mechanism, Integrated management services, Community culture

Table 2.4 The metrics of the service part for a complete community

Categories	Object	Area (m^2)	Total area	Density
Public service facility	A comprehensive service station	800	5070 m^2	≥3%
	A kindergarten	3500		
	A Nursery	300		
	An elderly service station	350		
	A community health station	120		
Convenient commercial services	A comprehensive supermarket	300	1700 m^2	≥1%
	A convenience store	300		
	Mail express delivery service	100		
	Convenient commercial outlets	1000		
Public activity space	Public activity place	150	4950 m^2	≥3%
	A multi-functional playground	800		
	A srden	4000		

2.5.3 15-Min Community-Based Life Range

Combined with the idea of a complete community is the 15-min community-based life range. It refers to the construction of a space circle where what is needed for living, working, travelling, and learning is within a 15-min walk [63]. Within this concept, the visualized realization of good education, good medicine for the sick, good care for the elderly, livable accommodation, and public support for the vulnerable groups are promoted by the actions of regeneration (Fig. 2.8).

The 15-min community life range represents both the hard power of a city and a symbol of its soft power. It embodies the quality of urban life through a healthy, comfortable, and civilized living environment, supported by comprehensive facilities and smart services for all ages. This includes nearby employment services, diverse community leisure spaces, an accessible learning system, a healthy life cycle with necessary services, and a strong sense of security for the elderly. Together, these features align with the concept of people-oriented urbanization [64]. The 15-min community emphasizes reasonable density, focusing on human-scale streets and the lived experience of residents [65]. It follows that urban planning should prioritize street-level experiences over rigid controls on building heights, and focus on the composition and diversity of urban functions rather than merely intensifying development.

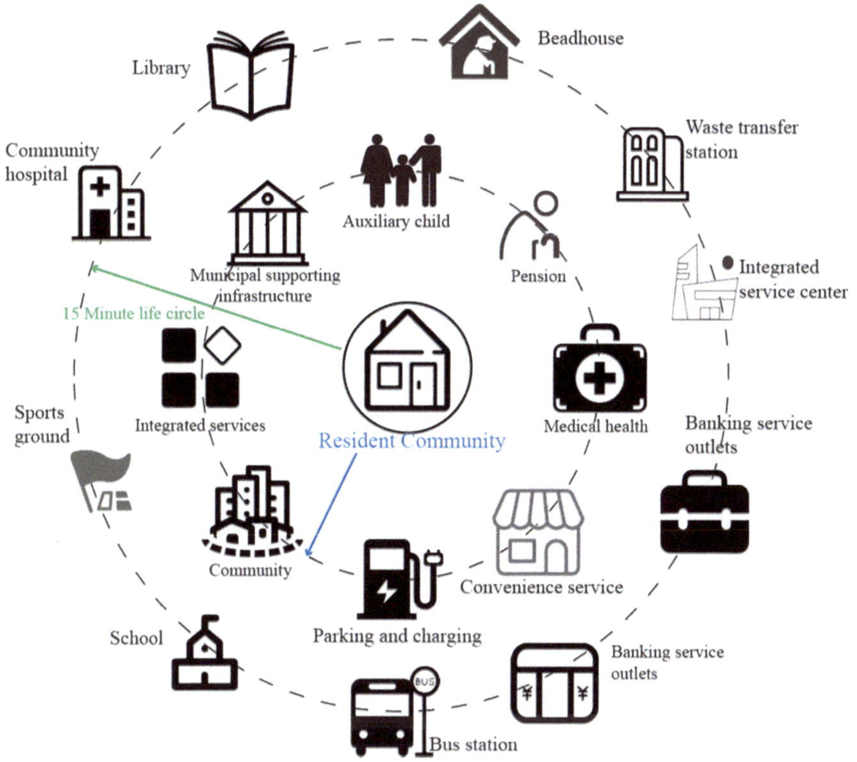

Fig. 2.8 Community-based life range in the complete community guideline (*Source* Image adapted by the authors from the Ministry of Housing and Urban–Rural Development (2021))

2.6 New Tools Toward the New Era: Evaluation Mechanism

In the new urbanization era, innovative tools must be introduced to transform old communities from a people-centered perspective. From an architectural point of view, the upgrading solutions should be diverse and adaptable. The primary focus of regeneration should be to enhance structural resilience while introducing multifunctionality (Fig. 2.9).

From an urban point of view, many regeneration and transformation methods have been tested in cities like Shanghai, Beijing, Guangzhou, and Shenzhen. These cases experimented with diverse designs, utilizing various densities and intensities of development strategies. The design process involved multi-channel approaches, integrating community planners and public participation mechanisms to bring together residents, community managers, investors, and government officials with designers. This collaborative process led to design proposals tailored to residents' needs, ultimately improving satisfaction within the communities.

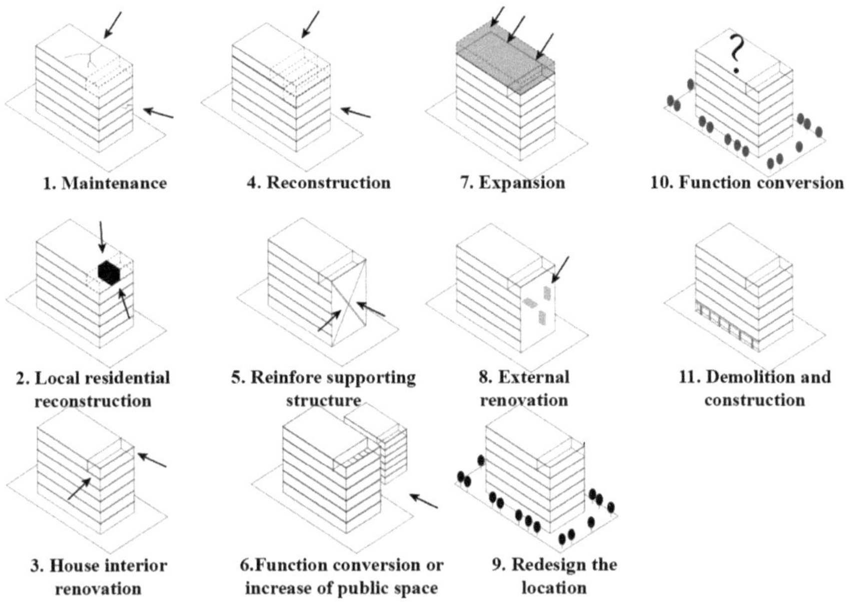

Fig. 2.9 The regeneration methods of existing multistorey houses (*Source* Image adapted by the authors from [56])

This participatory approach is rooted in China's National Territory Spatial Planning (NTSP), which aims to achieve sustainable development through a "multiple regulations in one" framework [66]. To enhance residents' perceptions and assist in design decision-making, systems like the "double-evaluation" and monitoring-evaluation-warning mechanisms have emerged within the NTSP. These focus on urban physical assessments at the macro level.

2.6.1 "Double Evaluation" in National Territory Spatial Planning

The "Double Evaluation" system is central to the Chinese spatial planning system (NTSP). It identifies conflicts and contradictions between different sub-categories of spatial plans and provides a comprehensive review of achievements and challenges. This system is essential for ensuring the effectiveness of planning, scientific index development, ecological sustainability, and development constraints. It also fosters collaboration between academia and government, promoting a shift from overall control to spatial guidance, which supports sustainable, high-quality, and coordinated development in the context of new urbanization. The double evaluation mechanism is divided into Current Situation Assessment and Planning Assessment [67].

2.6.1.1 Current Situation Assessment

The purpose is to improve the efficiency and precision of spatial governance, ensuring compliance with local needs. The assessment generates real-time dynamic feedback to identify issues and contradictions in land protection and development. The evaluation data, which should span multiple years, relies on diverse data sources, including social big data. New technologies such as Weibo Check-ins, POI, and mobile signaling also assist in data collection [68].

The evaluation system should be scientifically robust, locally relevant, and easy to implement. It should reflect local characteristics while being comprehensive, effective, and operational. The assessment process should incorporate multiple dimensions—target, region, time, and space—while comparing data with other similar cities over time. Results should be updated dynamically and integrated with land-use and spatial planning.

2.6.1.2 Planning Assessment

This focuses on evaluating various spatial planning approaches through a multi-perspective review, assessing urban master plans, land-use plans, and other significant spatial strategies. The evaluation includes three stages: Before, Ongoing, and After implementation. This assessment goes beyond traditional blueprint-based evaluations, incorporating indicator-based analysis and shifting from a static plan to a more flexible, target-oriented approach. The problem-oriented evaluation identifies implementation gaps by cross-examining existing spatial plans, highlighting discrepancies, and learning from past experiences [69]. Meanwhile, the target-oriented approach provides a clear direction for future planning by offering recommendations on spatial layout, indicators, and projects, laying the groundwork for effective monitoring. Public participation in planning evaluation enhances the fairness, rationality, and transparency of the process.

2.6.2 "Programming & Post-Occupancy Evaluation" in Architectural Design

The double evaluation system could be synthesized as the "Programming & Post-Occupancy Evaluation" mechanism [70]. This mechanism enables a controlled, dynamic process of design correction, prediction, and risk assessment through data analysis, decision-making technologies, and design feedback. Programming refers to analyzing the interaction between the building environment and its occupants. This includes overall planning and architectural design, guiding the integration of function and space [71]. Post-Occupancy Evaluation involves reviewing the project after completion to assess service conditions and overall design effectiveness.

The key technology of the "Programming & Post-Occupancy Evaluation" theory is space planning technology. Space planning technology mainly includes space generation, task evaluation, fuzzy preference decision-making, and other technical points.

- **Space generation**: Using simulation and numerical methods to establish relationships between design elements and optimizing spatial allocation.
- **Task evaluation**: Applying multi-variable analysis and spatial hierarchical processes to refine functional space allocation, identify potential risks, and reduce subjective biases.
- **Fuzzy preference decision:** Using demand-driven fuzzy logic to ensure objective, data-backed decision-making, eliminating irrational subjective judgments.

By combining the double evaluation system in NTSP with Programming & Post-Occupancy Evaluation, guidelines for old community regeneration can be developed, focusing on pre-residents' satisfaction evaluation and post-spatial quantitative analysis. This integrated approach offers a comprehensive framework for urban regeneration decision-making.

2.7 Conclusion

Understanding China's urbanization and identifying the "new" trends is crucial for shaping policy and urban development. As large cities experience population growth and face limited construction land due to central government constraints, the challenge becomes clear: how to accommodate more people within existing urban environments. This highlights the need for research into the contradiction between population growth and land scarcity.

Often urbanization in China prioritized rapid development and GDP growth, often at the expense of high-quality, people-oriented, sustainable development. To correct this course, the NTSP system emerged, introducing a "double evaluation". Process aimed at guiding planning decisions. This mechanism focuses on both situational and outcome assessments, ensuring long-term, comprehensive evaluation of the planning process and its implementation [72].

The "Programming & Post-Occupancy Evaluation" system has similarly gained traction, helping to shape decision-making based on real-world outcomes. Both systems share a people-oriented approach, emphasizing the needs and satisfaction of residents. By testing and evaluating various proposals, these mechanisms now serve as guiding frameworks for urban regeneration, especially at the community level in China.

However, more research is needed to further explore the relationship between these concepts. The following chapter will examine the discourse of density and densification, and investigate how these factors influence sustainable urban regeneration in China's aging communities.

References

1. Wu F (2022) Creating Chinese urbanism: urban revolution and governance changes. UCL Press. ISBN 1800083335
2. State Council of the People's Republic of China (2021) Outline of the 14th Five-Year Plan (2021–2025) for national economic and social development and long-range objectives through the year 2035 [R]. Central Party Literature Publishing House. Translated by CCPC Party History and Literature Research Institute (Official English version)
3. Xiang P, Wang Y, Deng Q (2017) Inclusive nature-based solutions for urban regeneration in a natural disaster vulnerability context: a case study of Chongqing, China. Sustainability 9:1205
4. Zhang X, Lu Y, Xu Y, Zhou C, Zou Y (2024) Governing regional inequality through regional cooperation? A case study of the Guangdong-Hong Kong-Macau Greater Bay Area. Appl Geogr 162:103135. https://doi.org/10.1016/j.apgeog.2023.103135
5. National Bureau of Statistics of China (2021, May 11) The seventh national population census bulletin (no. 5): age composition of the population [R].https://www.stats.gov.cn/. Accessed 30 June 2025
6. Zhang X, Zhou Z, Song Y (2024) Contested geopolitics of distribution in the city-region-building: promoting enclave industrial parks in the Pearl River Delta, China. Trans Inst Br Geogr 50(2):e12697
7. National Bureau of Statistics of China (2018) Sustainable and rapid development of construction industry and significant improvement of urban and rural landscape—series of reports on the achievements of economic and social development in the 70th anniversary of the founding of New China [R]. China Statistics Press, Beijing. http://www.stats.gov.cn/ztjc/zt70zn/. Accessed 30 June 2025
8. Zhao X, Chen J, Li J, Wang H, Zhang X, Yu F (2025) Unraveling the renewal priority of urban heritage communities via macro-micro dimensional assessment—a case study of Nanjing City, China. Sustain Cities Soc 124:106317
9. Song H, Chen J, Li P (2025) Decoding the cultural heritage tourism landscape and visitor crowding behavior from the multidimensional embodied perspective: insights from Chinese classical gardens. Tour Manag 110:105180. https://doi.org/10.1016/j.tourman.2025.105180
10. Chi W, Lin G (2019) The use of community greenways: a case study on a linear greenway space in high dense residential areas, Guangzhou. Land 8:188
11. Chen J, Li P, Wang H (2025) Assessment of influential factors on commute and life satisfaction in a historic campus-adjacent environment: evidence from a comparison study of twin cities. J Urban Plan Dev 151:4024065. https://doi.org/10.1061/JUPDDM.UPENG-5171
12. Chen J, Zhao X, Wang H, Yan J, Yang D, Xie K (2024) Portraying heritage corridor dynamics and cultivating conservation strategies based on environment spatial model: an integration of multi-source data and image semantic segmentation. Herit Sci 12:419. https://doi.org/10.1186/s40494-024-01497-7
13. Zhang X, Sun Y, Luan X (2023) Regulation flexibility and legitimacy building in governing intercity railways: the polymorphous role of the Chinese provincial government. J Urban Aff 1–16
14. People's Daily (2019, Sept 25) China builds the world's largest housing security system—striving to achieve all people's housing [N]. People's Daily Publishing House, Beijing. http://en.people.cn. Accessed 30 June 2025
15. Morgan Stanley (2021) China's smarter cities: the key to growth [R]. Morgan Stanley Research, New York. https://www.morganstanley.com. Accessed 30 June 2025
16. Li Y, Zhang X, Xia C (2023) Towards a greening city: how does regional cooperation promote urban green space in the Guangdong-Hong Kong-Macau Greater Bay Area? Urban For Urban Green 86:128033
17. Zhang X, Shen J, Sun Y, Zhou C, Yang Y (2023) Formation of city regions from bottom-up initiatives: investigating coalitional developmentalism in the Pearl River Delta. Ann Am Assoc Geogr 113:700–716

18. The State Council. The Thirteenth Five-Year Planf China
19. Pan L, Gan W, Chen J, Xian X (2024) Ecological planning for mitigating urban heat based on thermal network simulation: a case study of Guangzhou-Foshan metropolitan area. J Chin Urban For 22:10–19. https://doi.org/10.12169/zgcsly.2024.06.26.0004
20. Zhang X, Chen J, Wang H, Yang D (2025) From policy synergy to equitable greenspace: unveiling the multifaceted effects of regional cooperation upon urban greenspace exposure inequality in China's megacity-regions. Appl Geogr 174:103472. https://doi.org/10.1016/j.apgeog.2024.103472
21. People's Daily (2020, Oct 10) During the 13th five-year period, China's Shantytown reform has started over 23 million units [N]. People's Daily Publishing House, Beijing. http://en.people.cn. Accessed 30 June 2025
22. Chen J, Ren K, Li P, Wang H, Zhou P (2024) Toward effective urban regeneration post-Covid-19: urban vitality assessment to evaluate people preferences and place settings integrating LBSNs and POI. Environ Dev Sustain. https://doi.org/10.1007/s10668-024-05378-8
23. State Council of China (2020, July 20) Guiding opinions on comprehensively promoting renovation of old urban communities [R]. State Council General Office, Beijing. http://www.gov.cn/zhengce/content/2020-07/20/content_5528320.htm. Accessed 30 June 2025
24. Zhao M, Liu N, Chen J, Wang D, Li P, Yang D, Zhou P (2024) Navigating post-Covid-19 social–spatial inequity: unravelling the nexus between community conditions, social perception, and spatial differentiation. Land 13(4):563
25. Liu C, Liu L, Shi B, Wen Y (2012) Existing problems and solutions of property management in old communities. Urban Probl 3
26. Wu Z, Wu J, Zhang J, Li H, Zhao Y, Yang J, Wang W, Zhou J, Liu J, Si M et al (2021) Symposium on implementation mechanism of renewing old urban quarters. Urban Plan Forum 10
27. Hu J, Chen J, Li P, Yan J, Wang H (2024) Systematic review of socially sustainable and community regeneration: research traits, focal points, and future trajectories. Buildings 14:881. https://doi.org/10.3390/buildings14040881
28. Chen J, Gan W, Liu N, Li P, Wang H, Zhao X, Yang D (2024) community quality evaluation for socially sustainable regeneration: a study using multi-sourced geospatial data and AI-based image semantic segmentation. ISPRS Int J Geo Inf 13. https://doi.org/10.3390/ijgi13050167
29. Lianxun Securities Research (2021) Old reform boosts domestic demand, opens trillion-yuan market [R]. Lianxun Securities, Shanghai (Industry Analysis Report)
30. Chen J, Li P, Wang H (2024) MSPA-based collaborative evaluation and planning governance of territory and natural environment in metropolitan agglomerations—a case study of EGIDDZ in Yangtze River Delta. J Resour Ecol 15:1263–1273. https://doi.org/10.5814/j.issn.1674-764x.2024.05.014
31. Alipui N, Gerke N (2018) The refugee crisis and the rights of children: perspectives on community-based resettlement programs. New Dir Child Adolesc Dev 2018:91–98
32. Leishman C, Rowley S (2012) Affordable housing. Sage Handb Hous Stud 379–396
33. Franklin B, Tait M (2002) Constructing an image: the urban village concept in the UK. Plan Theory 1:250–272
34. Li H, Hu J (2019) A study on classification and governance of Chinese communities. *Macroeconomics* 15
35. Zhang X, Chen Y, Xu Y, Yue W, Tang C (2023) Does regional cooperation constrain urban sprawl? Evidence from the Guangdong-Hong Kong-Macao Greater Bay Area. Landsc Urban Plan 235:104742
36. Liu J, Zhang Y, Ran A (2020) Research on the renewal and transformation of old districts in Beijing: based on characteristics-dilemma-policy analysis framework. Soc Gov Rev 10
37. Chen J, Pellegrini P, Yang Z, Wang H (2023) Strategies for sustainable urban renewal: community-scale GIS-based analysis for densification decision making. Sustainability 15:7901
38. Chen J, Li P, Lei Y, Zhang Y, Lai C, Chen B, Liu J, Schnabel MA (2025) Leveraging augmented reality for historic streetscape regeneration decision making: a big and small data approach with social media and stakeholder participation integration. Cities. https://doi.org/10.1016/j.cities.2025.106214

39. Chen X, Duan J (2022) What they talk about when they talk about urban regeneration: understanding the concept 'urban regeneration' in PRD, China. Cities 130:103880
40. Zhong X, Chen X (2017) Demolition, rehabilitation, and conservation: heritage in Shanghai's urban regeneration, 1990–2015. J Archit Urban 41:82–91
41. Yuan Y, Song W (2020) Mechanism and effect of shantytown reconstruction under balanced and full development: a case study of Nanjing, China. Sustainability 12:7979
42. Guangdong Provincial Government (2021) Three olds reform policy [EB/OL]. https://www.gd.gov.cn/zwgk/wjk/qbwj/yf/content/post_2595862.html. Accessed 30 June 2025
43. Shenzhen Municipal Government (2021, March 30) Measures for urban renewal of Shenzhen [EB/OL]. https://sf.sz.gov.cn/fggzywyb. Accessed 30 June 2025
44. Shenzhen Bureau of Planning and Natural Resources (2010) Shenzhen urban master plan (2010–2020) [R]. Official Gazette, Shenzhen
45. Central Committee of CPC (2013, Dec 12) Report of the central conference on urbanization [R]. People's Publishing House, Beijing
46. Guangzhou Bureau of Planning and Natural Resources (2020) Circular on strengthening control of farmland protection systems [R]. Official Gazette, Guangzhou
47. State Council of China (2013) National new urbanization plan (2014–2020) [R]. People's Publishing House, Beijing. ISBN 978-7-01-013586-1
48. State Council General Office (2016, June 3) Opinions on accelerating development of housing rental markets [R]. State Council. http://www.gov.cn/zhengce/content/2016-06/03/content_5079330.htm. Accessed 30 June 2025
49. Shanghai Natural Resources and Planning Bureau (2021) Detailed rules for implementing urban renewal planning in Shanghai [R]. Shanghai Gazette, Shanghai
50. Xinhua News Agency (2020, Dec 17) Communiqué of the central economic work conference [N]. Xinhua, Beijing
51. Ren X (2013) Urban China [M]. John Wiley & Sons, Hoboken. ISBN 0745665454
52. Ma G, Paola P, Wu H, Han H, Wang D, Chen J (2023) Impact of land-use mixing on the vitality of urban parks: evidence from big data analysis in Suzhou, Yangtze River Delta Region, China. J Urban Plan Dev 149:4023045. https://doi.org/10.1061/JUPDDM.UPENG-4334
53. Power A (2010) Housing and sustainability: demolition or refurbishment? Proc Inst Civ Eng Des Plan 163:205–216
54. Ministry of Housing and Urban-Rural Development (2021, Aug 10) Circular on Preventing Large-Scale Demolition in Urban Regeneration [R]. MHURD, Beijing. http://www.mohurd.gov.cn/gongkai/fdzdgknr/zfhcxjsbwj/202108/20210810_251161.html. Accessed: 30 June 2025
55. Chen M, Ye C, Lu D, Sui Y, Guo S (2019) Cognition and construction of the theoretical connotations of new urbanization with Chinese characteristics. J Geogr Sci 29:1681–1698
56. Matsumura S, Yue F, Tongtong L (2008) Reborn European American collective housing. China Machine Press
57. Wu X, Wang Z (2022) Multi-objective optimal allocation of regional water resources based on slime mould algorithm. J Supercomput 1–30
58. Liu Y, Zhu A-X, Wang J, Li W, Hu G, Hu Y (2019) Land-use decision support in brownfield redevelopment for urban renewal based on crowdsourced data and a presence-and-background learning (PBL) method. Land Use Policy 88:104188
59. Ministry of Housing and Urban-Rural Development (2021) Complete residential community construction guide [R]. Ministry of Housing and Urban-Rural Development, Beijing (Document No.: *Jian Ban Ke* [2021] No. 55). https://www.mohurd.gov.cn. Accessed 30 June 2025
60. Liu G, Chen S, Gu J (2019) Urban renewal simulation with spatial, economic and policy dynamics: the rent-gap theory-based model and the case study of Chongqing. Land Use Policy 86:238–252
61. Chen J, Tian W, Xu K, Pellegrini P (2022) Testing small-scale vitality measurement based on 5D model assessment with multi-source data: a resettlement community case in Suzhou. ISPRS Int J Geo Inf 11:626. https://doi.org/10.3390/ijgi11120626

62. Groulx M, Nowak N, Levy K, Booth A (2021) Community needs and interests in university-community partnerships for sustainable development. Int J Sustain High Educ 22:274–290
63. Pozoukidou G, Chatziyiannaki Z (2021) 15-Minute city: decomposing the new urban planning Eutopia. Sustainability 13:928
64. Guzman LA, Arellana J, Oviedo D, Aristizábal CAM (2021) COVID-19, activity and mobility patterns in Bogotá. Are we ready for a '15-minute city'? Travel Behav Soc 24:245–256
65. Wu H, Wang L, Zhang Z, Gao J (2021) Analysis and optimization of 15-minute community life circle based on supply and demand matching: a case study of Shanghai. PLoS ONE 16:e0256904
66. Song R, Hu Y, Li M (2021) Chinese pattern of urban development quality assessment: a perspective based on national territory spatial planning initiatives. Land 10:12–15
67. Zou R, Kang J (2014) Thoughts on urban renewal based on the background of "three old" transformation—taking the "three old" transformation plan of Qiaokou District of Wuhan as an example. In Proceedings of the 2014 China Urban Planning Annual Meeting Papers
68. Jiang Y (2021) A study on urban vitality measurement and its spatial characteristics—a case study of the main urban area of Wuhan. Urban Manag Sci Technol 22:4
69. Long Y, Huang CC (2019) Does block size matter? The impact of urban design on economic vitality for Chinese cities. Environ Plan B Urban Anal City Sci 46:406–422
70. Zhuang W (2017) From architectural programming to post occupancy evaluation: a feedback mechanism of architectural procedure circle. Des Community 5:125–129
71. Zhao X, Yu F, Zhang X, Chen J, Li P (2025) Assessing urban renewal efficiency via multi-source data and did-based comparison between historical districts. npj Herit Sci 13:389. https://doi.org/10.1038/s40494-025-01964-9
72. Pellegrini P, Chen J (2025) The regeneration path not taken further: an experiment in urban densification and state entrepreneurialism in a resettlement neighborhood in Suzhou. China Urban Stud 00420980251365188. https://doi.org/10.1177/00420980251365188

Chapter 3
Density and Densification: An Evergreen Issue

3.1 Introduction

Density is a key parameter of the built environment, profoundly influencing urban form and functionality [1, 2]. Its discussion is crucial in defining sustainable community regeneration and informing urban design strategies [3, 4].

J. Chen and P. Pellegrini, *Densification as a Design Tool for Sustainable Urban Regeneration*, Urban Sustainability, https://doi.org/10.1007/978-981-95-1653-7_3

This chapter introduces the definition and measurement methods of high-density communities in China, integrating relevant theoretical research on urban density. It offers a multi-dimensional exploration of how density is perceived and understood in various contexts [5, 6]. Furthermore, it examines the relationship between densification and sustainable regeneration, proposing the concept of "moderate density" as a flexible solution for urban regeneration within China's ongoing urbanization. This concept is adaptable to various contexts, conditions, and resource availability, avoiding extreme approaches [7, 8]. The term "moderate" reflects a growing discourse in China, emphasizing a balance between development intensity, diversified open spaces, and multi-level urban interfaces, while steering clear of both hyper-dense and under-utilized urban spaces. Moderate density is suggested as a guiding principle for decision-making in regeneration processes [9, 10].

3.2 High-Density in China

3.2.1 Density Measurement

Density is commonly expressed using various metrics to quantify urban space, such as population density, construction density, and employment density [11]. Population density, often defined as the number of people per unit area, can also be detailed in more technical terms, such as floor area ratio (FAR) or distinctions between gross density and net density [12]. Methods of calculating density are multiple and vary by region, reflecting differing urban types, cultural contexts, and governmental definitions [13]. In fact, demographic variables as well as geo-metrics indicators vary worldwide in different urban types, societies, and cultures (e.g., number of inhabitants per household); different governments have different definitions of density related to the measures [14]. Take Hong Kong as an example, if the land area of the whole territory is considered, the overall population density in Hong Kong is about 6300 people per sq. km. However, only about 24% of the total area in Hong Kong is built up. Therefore, if the geographical reference is confined to built-up land, the population density will be about 25,900 people per sq. km, four times the territory's overall density [15].

There is usually a difference between China and Europe in the calculation of building density, which is related to the way the building process is governed and managed [16]. In Europe, due to the private ownership of land, usually the *net density* is calculated: density is within the allotment, and elements outside the property, such as roads and green spaces, are not considered [17]. The calculation in China instead serves the comprehensive development process, which almost always is applied to a large portion of land which belongs to the state. Therefore, it is calculated as the gross density, and includes parcels, roads, green spaces, parking, and other infrastructures [18]. In addition, population density in China is often assessed at the regional or community level, integrating residential elements, public facilities, infrastructure, and natural features into the measurement.

Some examples can be made. In the UK, the main measurements of physical density are divided into three categories:

1. Net density: habitable rooms per hectare (hr/ha).
2. Gross density: similar to net density but includes other land uses such as parks, schools, and roads.
3. Plot ratio (FAR): the ratio of the total built floor area to the site area [14].

In China, since land ownership belongs to the state, the physical density is usually measured in two parts.

1. Architectural density: the ratio of building structures to the land area.
2. FAR: the total gross floor area relative to the site area.

Density and its measurement are complex and sometimes controversial. For instance, while some studies suggest China is densifying its urban areas, others report a decrease in density [19]. Both interpretations are valid, depending on whether the calculations consider the entire metropolitan area or select urban districts. In cities like Shanghai, for example, within a 10 km radius of the center, population density reaches 25,600 people per square kilometer—almost double that of Tokyo or New York. In Suzhou, although the overall population density is lower, the Gusu district sees a density of 11,076 people per square kilometer—nearly seven times the city's average [20] (Table 3.1). In contrast, Wuzhong, Xiangcheng, and Wujiang districts have lower densities, due to large farmland and water areas. In such cases, focusing the calculation on built-up land offers a clearer picture of urban density.

Population density varies across different urban agglomeration forms, such as decentralized and centralized patterns. As illustrated in Fig. 3.1, the decline in decentralized population density from the center to the edge is non-linear and random, while centralized population density decreases more uniformly from the core to the periphery [15].

Density also differs depending on individual perception. In any given urban environment, physical density (the actual number of people per unit area) and perceived

Table 3.1 Metropolitan Suzhou population density in 2021

Districts	2021 (million)	Aera (km^2)	Pops density
Suzhou (6 districts)	6.7148	2899.38	2315
Gusu	0.924	83.42	11,076
Wuzhong	1.3889	745	1864
Xiangcheng	0.891	489.96	1818
SND - Suzhou New District	0.832	332.37	2503
SIP - Suzhou Industrial Park	1.1339	278.19	4076
Wujiang	1.545	970.44	1592

Source Suzhou Municipal Bureau Statistics (2021)

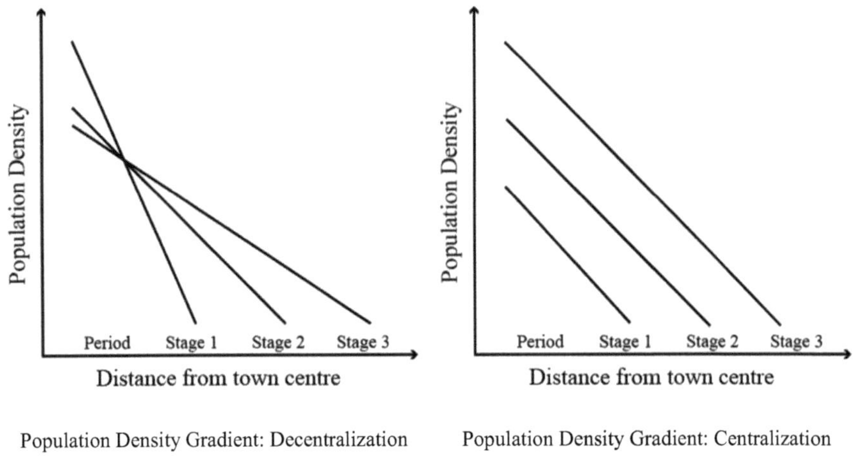

Fig. 3.1 Population density gradient categories (*Source* Image adapted by the authors from [22, 23])

density (how dense people feel an area is) can diverge significantly [21]. People often develop distinct perceptions of density in specific areas, shaped by various indicators, and synthesize the relationship between physical and perceived density.

In summary, density figures fluctuate greatly based on perception and calculation methods, with each approach being appropriate for different contexts. This study primarily uses the gross density calculation method. For the topic of "high-density communities," the central debate revolves around defining high-density and ultra-high-density [22, 23].

3.2.2 High-Density Definition and Metrics

The concept of building or population density remains an issue deeply rooted in the specific context. It encompasses not only the geographical environment but also the cultural and social contexts, with no consensus on the defining thresholds [24]. Similarly, the debate over what constitutes an ideal density level for urban development persists. The interpretation of high-density development is equally controversial.

Key factors influencing high-density urban development include population size, land area, urbanization stage, industrialization level, resources, culture, religion, and political systems. High-density environments are typically characterized by the following elements: (1) high building floor area ratio; (2) high building coverage; (3) low open space ratio; and (4) large amount of population.

Scholars have attempted to define high density in various ways. A commonly accepted threshold defines high-density areas as those with a plot ratio exceeding 1.6, and a population density greater than 15,000 people per square kilometer (see

Table 3.2 Density definition with FAR and population density

Type	Feature	Low	Medium	High	Ultra-high
Building floor area ratio (FAR)	Building capacity intensity index	<0.75	0.75 – 1.60	**>1.60**	
Population density	Sample threshold value of urban blocks			15,000 pop. /km^2	25,000 pop. /km^2

Source According to the study and analysis of Li & Ye, 2015 [28]

Table 3.2). Other scholars, after reviewing existing literature, suggest that ultra-high-density cities should have a population density above 25,000 people per square kilometer [25]. According to this criterion, 76 cities worldwide qualify as high-density, predominantly in developing countries with low urbanization levels and industrialization [26, 27].

Rankings from authoritative institutions, such as the World Atlas, UN-Habitat, Demographia World Urban Areas, the Atlas of Urban Expansion, and the Global City Power Index,[1] often disagree on the definition. Typically, high-density refers to a population density of around 10,000 people per square kilometer. However, the 15,000 people per square kilometer threshold proposed by Li and Ye fails to reflect various scenarios in China. If population density alone serves as the benchmark for determining development intensity, then most regions in China would fall under high-density or even ultra-high-density categories.

These divergent interpretations highlight that density standards must be viewed in a more comprehensive context, rather than being based solely on numerical thresholds. Theoretical discussion is crucial to better understand the underlying meanings of density. The next section explores the theoretical evolution of density and examines the advantages and challenges of high-density development [29–31].

3.3 Review of Density Development

3.3.1 Main Ideas About Urban Density Levels

To address the challenges brought on by industrialization and urbanization, scholars have gradually developed two models to assess urban density and its implications: decentralization with polycentrism and centralization [22] (Fig. 3.2). The centralization model defines a hierarchical system with a central node, where all other nodes depend on and are connected to this central point. In contrast, decentralization with polycentrism describes a structure in which multiple centers are linked to higher-level nodes and other connecting nodes [32].

---

[1] Available online: Demographia World Urban Areas 18th Annual Edition 2022, http://www.demographia.com.

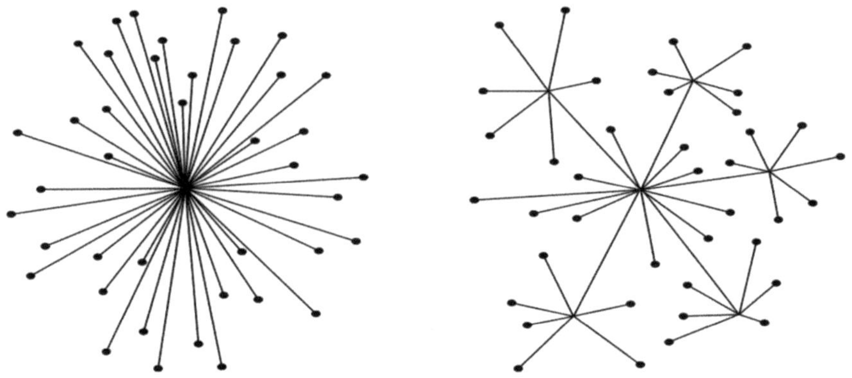

Fig. 3.2 The space pattern centralization (*left*) and polycentrism (*right*) (*Source* Image adapted by the authors from [33])

These two approaches often define specific density levels for various elements. A key example of decentralization is Howard's Garden City idea, which advocates for controlling urban population density within 8000 people per sq. km, allocating 1000 acres of urban land (approximately 4.05 km^2), 5,000 acres of agricultural land, and a maximum of 32,000 people.

Centralization, or the proposal to concentrate residents and activities, is epitomized in Corbusier's La Ville Radieuse (1935), where high-density living is used to address urban challenges such as rapid population growth, poor sanitation, and traffic congestion [34]. Corbusier's projects proposed increasing building heights, expanding green spaces, and widening roads. In La Ville Radieuse, high-rise buildings could house up to 50,000 people, and the population density could reach nearly 80,000 people per sq. km.

Scholars have fiercely debated the merits of these approaches. Some, like Louis Wirth in his "Urbanization as a Way of Life" [35], strongly advocated for high-density urban living. Wirth defined a city as "a relatively large, dense, and permanent settlement of socially heterogeneous individuals" [36], placing "density" itself at the core of the definition of a city.

Ludwig Hilberseimer, in his work Metropolis-Architecture, envisioned the modern city as a superposition of two types of spaces: residential units stacked above office and workshop spaces. This high-rise model combined living, recreation, and work functions, reflecting the era's emphasis on functional urban design for modern society [37, 38].

On the other hand, many scholars recognized the negative impacts of high-density development on urban society and the environment. Eliel Saarinen, in *The City: Its Growth, Its Decay, Its Future*, proposed the theory of Organic Decentralization as a response to over-concentration [39]. Saarinen argued for moving heavy and light industries out of city centers, freeing up space to enhance green areas and improve living conditions. He advocated for centralizing living and working areas while decentralizing "occasional activities" [40].

When it comes to high-density, high-vitality urban environments, Jane Jacobs sharply criticized functionalist planning yet embraced high-density. She supported urban vitality through short streets, pedestrian density, mixed land uses, and a diversity of building types [41]. Jacobs believed these elements would naturally support urban diversity and vitality. Similarly, Jan Gehl emphasized four principles of urban space design: integrate or aggregate, assemble or separate, invite or reply, and open up or close in. These principles suggest that cities should be multifaceted, not monolithic [42].

In practice, policies such as zoning—the primary tool for land management in the U.S.—tend to offer a simplified approach, advocating for multi-density development rather than one singular solution. The *Standard State Zoning Enabling Act* divides urban land into residential, commercial, and industrial zones, each further categorized into low, medium, and high-density areas [43]. This approach allows for flexibility, enabling land-use modifications to meet the diverse needs of urban residents.

In the majority of recent interpretations, high-density is always combined with compactness, and theories like "Compact City" [44], "New Urbanism" [45], "Smart Growth" [46] advocate for land efficiency, reduced energy consumption, and urban revitalization through denser development. Although interpretations of compact cities vary, they generally emphasize high-density, mixed land use, and transit-oriented development (TOD). As a result, urban compactness has become a critical evaluation metric for high-density cities [47]. European countries widely support compact cities due to their high energy efficiency, economic development potential, and diverse urban vitality, making them the prevailing model of urban development today [48].

3.3.2 High-Density Development

Since the turn of the twenty-first century, the rise of high-density cities has been driven largely by capital forces, resulting in rapid urbanization across the globe [44]. However, this intense urban development has led to significant challenges. Overcrowded urban centers suffer from issues like traffic congestion, declining spatial quality, a shortage of green spaces, severe air pollution, and the degradation of the ecological environment [49].

In the ongoing debate surrounding urban density, scholars have examined both the positive and negative effects of high-density development from various perspectives. From an economic standpoint, research has demonstrated an inverted U-shaped relationship between urban density and economic output efficiency, based on empirical data. At low density levels, increasing density can reduce land consumption and infrastructure costs, driving economic growth. However, once density surpasses a critical threshold, the negative impacts—such as traffic congestion, air and noise pollution—outweigh the economic benefits, ultimately hindering further growth [50, 51]. The key challenge, therefore, is defining this critical threshold.

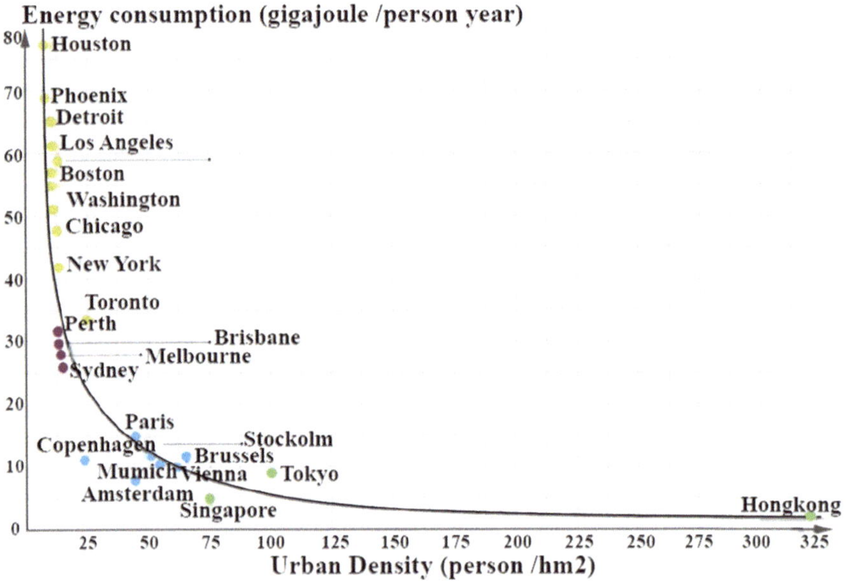

Fig. 3.3 Relationship between urban density and traffic energy consumption (*Source* Image adapted by the authors from [52])

From an energy consumption perspective, higher urban density has been shown to enhance land use and improve public transport efficiency, ultimately reducing urban energy consumption (Fig. 3.3) [53]. For example, megacities in Asia, such as Hong Kong, Singapore, and Tokyo, exhibit much lower per capita transportation energy consumption than major cities in Australia and North America.

In terms of transportation energy consumption and carbon emissions, increasing population and building density can reduce transportation energy consumption [54]. However, when density exceeds the carrying capacity of public transport, it can lead to higher air pollution [55]. A study of 10 central stations in Shanghai found that the sustainable development intensity around stations, influenced by varying daily passenger circulation, should prioritize actual usage when defining density [56]. For small and medium-sized stations (≤50,000 people), the FAR should be controlled between 1.5 and 2.0; for larger stations (50,000 to 100,000 people), between 2.0 and 2.5; and for super-large stations (≥100,000 people), between 2.5 and 3.0.

From a building energy consumption perspective, studies suggest that the relationship between building height, density, and energy consumption follows a U-shaped curve (Fig. 3.4) [57]. Analysis of numerous neighborhoods in Portland and Atlanta reveals this U-shape between building height and heating and cooling energy consumption per unit area under various building densities. The lowest energy consumption occurs when building height is around six floors [52]. When building height remains constant, higher building density leads to slightly lower

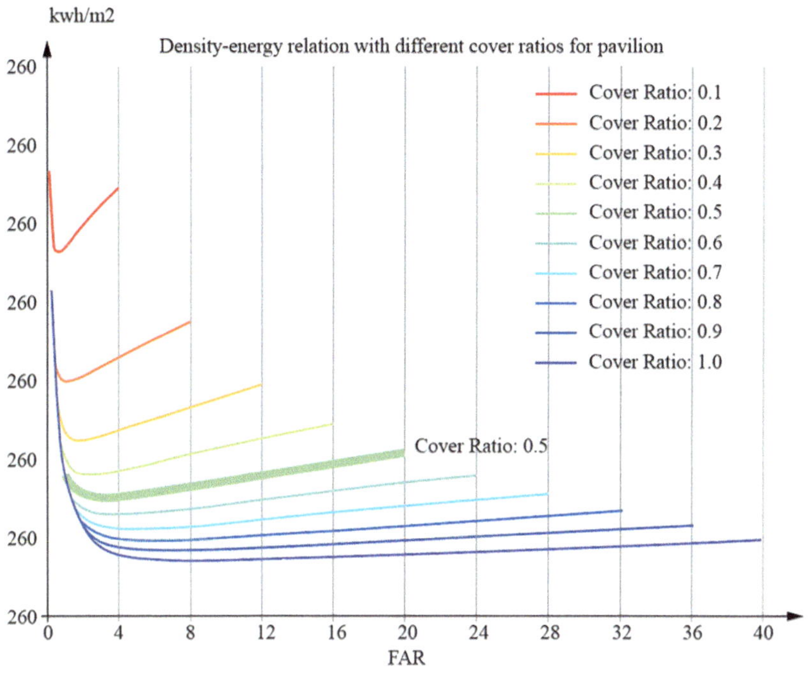

Fig. 3.4 Density-energy relation with different cover ratios (*Source* Image adapted by the authors from [52])

energy consumption per unit area for heating and cooling, though the overall impact remains modest.

In summary, mainstream research on the density of the built environment typically focuses on traffic capacity, urban space distribution, and economic development. Recent studies emphasize that high-density development should consider agglomeration, reduce energy consumption, and balance spatial quality trade-offs, all while moderately improving urban quality at higher densities [58].

3.4 Density and Sustainability

According to the majority of studies and policies in city planning density plays an essential role in achieving sustainable development [59]. Many scholars explored the relationship between density and sustainability, and high density has the advantage of efficient land and resource utilization [60]. In October 2016, the United Nations adopted the Action Plan on Housing and Sustainable Urban Development at the Habitat III conference in Quito, Ecuador, promoting a policy of "utilization density and economies of scale." This plan aims to integrate environmental, social, and economic agendas [61]. The number of residents per land unit is closely linked to land consumption, which should be minimized [62].

Density is intrinsically connected to urban form. Its interaction with factors such as infrastructure, open space, and design quality influences urban morphology [63]. Infrastructure availability, environmental and heritage policies, and building regulations all shape density levels [64]. According to the UN-Habitat Principles for Sustainable Neighborhoods, the performance guide advocates for high-level amenities in city centers and target areas based on a vision of dense, mixed-use, mixed-income zones with concentrated public transport. By pooling human, material, and financial resources, dense settlements are intrinsically linked to efficient, inclusive, and flexible development forms. From an economic standpoint, the focus has shifted from dense and compact cities' traditional environmental and social roles to their ability to drive smart growth and long-term progress [65].

Proponents of high-density development argue that densification is an inclusive process. It can create opportunities for urban residents [66], because it can boost the economy, generate jobs through economies of scale, increase social vitality, improve access to urban services, and reduce per capita resource consumption and greenhouse gas emissions to combat climate change [67] (Table 3.3). The inclusiveness of densification largely depends on how it is implemented within communities.

However, the economic pressures associated with densification can drive up real estate prices, rather than increase available housing. This may result in the displacement of existing residents and companies, worsening the housing affordability crisis in some cities. As middle-income groups seek proximity to the city center amidst deteriorating traffic conditions, the real estate market may adapt by crowding more people into the same properties, often bypassing space standards [68].

In the context of the 2019 coronavirus disease pandemic, how to solve the informal densification process became particularly prominent. In the dense spaces, the required social distance and driving away from the public space is was almost impossible, and the service level may not be enough to protect public health [74]. The problem is not the densification, but the architectural form and socio-economic environment involved. Without effective regulators and investment in supporting infrastructure, the concentration of entities driven by desperate or narrow business computing can have devastating consequences.

The government plays a crucial role in guiding urban development and preventing uncontrolled growth [75, 76]. This requires the leadership to have a clear vision of the future form of the city and to recognize the importance of intensive activities in appropriate places, which are surrounded by a variety of communities with easy access to work and facilities.

To sustain the densification process, it needs to enlist community support for the encryption initiative and coordinate public and private investment flows to make it a reality. A series of formal regulations and positive measures are needed to ensure the efficient, orderly, and safe development and minimize the risks involved [77].

Table 3.3 High density debate: key potential advantages and disadvantages of intensification

	Advantages	Disadvantages
Environmental sustainability	Reducing private car use	Increase in traffic in town centers
	Promoting the use of public transport, walking and cycling	Heat island effect in highly dense urban areas
	Reducing the need for the development of existing green fields	
	Reducing heat loss and energy consumption in buildings	
	Reducing carbon dioxide emission	
	Reducing energy consumption	
Social sustainability	Developing active and vibrant public spaces	Overcrowded neighborhoods
	Developing a sense of community	Crime rate and safety issues in high-density neighborhoods
		A negative perception of high-density urban areas
Economic sustainability	Improving social equity via reducing the cost of transport and the need for private cars	The threat of lacking appropriate facilities and services for the large population
	Potential for mixed-use development	Increase of density without consideration for jobs and the housing market
	Viability of investment in public transport	
	Viability of new local jobs	

Source Image adapted by the authors and based on [69–73]

3.4.1 *Positive Elements of High-Density for Sustainability*

According to Boyko and Cooper, in high-density built environment, there are ***social opportunities and diversity advantage*** [60], and other positive conditions and effects as follows:

A high-density built environment may enhance social support, social capital, and attachment to community, thus reducing social isolation, exclusion, and segregation.

A high-density built environment can increase diversity, security, vitality, and opportunities for creativity and social interaction.

A high-density built environment can create a more livable and sustainable urban environment, promote more activities and vitality in the city center.

Recent economic studies suggest that density generates positive value and efficiency in production and consumption activities [78]. These advantages stem from the "city premium", where proximity enables task concentration, improves expertise, and fosters synergy through complementary functions. These interactions create "positive externalities" and social interaction, driving creativity, talent attraction, and accelerated investment and employment [79].

Moreover, high-density environments offer green space, physical benefits (e.g., street visibility), and energy savings. A compact urban form allows for larger public open spaces and reduces traffic and energy costs. The United Nations' New Urban Agenda (NUA) advocates urban concentration over sprawl, emphasizing agglomeration economies, resource efficiency, and reduced mobility and public service costs. These principles aim to support social inclusion by ensuring affordable housing in mixed-income, mixed-use areas.

3.4.2 Negatives Elements of High-Density for Sustainability

Despite its advantages, high density can lead to traffic congestion, strain on public infrastructure, poor public health, and rising housing prices if urban morphology, infrastructure, and governance are poorly designed and managed [26, 37].

Further, high-density areas may overlook people's perceptions and the trade-offs between density and various urban issues. A lack of clarity in align in residents' needs with policy-making can lead to misguided density decisions based on unreliable data. This misalignment may affect land value and produce unintended outcomes. For example, including a public park in density calculations may alter the proposed neighborhood density and impact housing development. Effective urban planning requires clear guidelines on density's role in environmental design and development [80]. However, policies must remain adaptable, as local factors will influence perceptions of density [81].

High density environments' contributions to well-being depend on a complex mix of institutions, investments, and infrastructure. In informal settlements, overcrowding often leads to poor access to essential services. Increased vehicles and population growth can worsen traffic congestion and air pollution [82]. Overburdened infrastructure—such as water, sanitation, and power systems—can result in bottlenecks and failures. These factors also reduce public open spaces, increasing social pressure and spreading diseases [83]. Additionally, high-density urban projects come with higher construction and maintenance costs compared to medium or low-density alternatives, raising prices for housing, goods, and services.

3.4.3 *Defining "Proper" Density*

The decision on density should integrate residents' perceptions, economic agglomeration, cultural aesthetics, and sustainable development goals. It's essential to consider perceived density, not just data-driven metrics [84]. Perceived density refers to how individuals perceive and estimate the number of people and activities in their environment [11].

In China, urban reforms have significantly improved living standards and reduced poverty [85]. However, the shift in urban forms—especially increasing population density—has not unequivocally improved the overall quality of life (QoL) [86]. The rapid transition from horizontal to vertical living has occurred in a single generation, requiring detailed local research to understand its consequences. Factors such as landscaping, building types, design, noise, and aesthetic qualities are critical in shaping the perceived density [87].

Urban economics links density increases to rising real estate prices, highlighting the relationship between access, price, and density [88]. China's land-driven urbanization has led to spatial separation between residential, commercial, and service areas, influencing density policies and real estate developers' strategies [89]. Culturally, some regions may regard the current density as approaching saturation. However, cultural shifts are ongoing. Cities worldwide, whether in the Global North or South, are embracing vertical expansion, driven by financial capitalism in urban development. Interestingly, many of these density increases are not directly tied to economic globalization [90].

A study used multivariate spatial data to analyze the relationship between population density, economic density, and economic growth across the globe [91]. It found that the optimal density for economic growth is around 20,000 people per sq.km, which corresponds to a plot ratio of 2.0 in residential and commercial areas in China. This moderate density is considered appropriate for sustainable development.

In sum, proper density integrates residents' perceptions, cultural preferences, economic dynamics, and environmental sustainability. Achieving the right density, through densification, can drive sustainable community regeneration.

3.5 Densification

3.5.1 *Densification Definition*

Densification is the process of intensifying various forms and levels of density through multiple strategies [92]. It transforms urban environments, enhancing vibrancy and activity in mixed-use spaces, maximizing the use of infrastructure, public facilities, and open spaces [73]. Densification has emerged as a central policy tool in efforts to realize sustainable urban forms such as smart growth, TOD, and New Urbanism [48]. These frameworks advocate for policies that capitalize on density and economies of scale to create synergies between the environment and economy, fostering sustainable development.

While densification offers a pragmatic solution to urban expansion, its benefits are often diluted by disordered growth and rapid development [93]. In China's urbanization process over the past 40 years, incremental planning has driven densification in suburban areas through reconstruction and new social housing projects [94]. However, sustainable densification often suffers from uncoordinated growth, highlighting the need for redefined policies to better integrate densification with the broader goals of urban sustainability [95].

3.5.2 *Current Density of Large and Medium-Sized Cities in China*

China's urbanization has long been characterized by rapid expansion and quantitative growth. As the urban–rural divide has shifted, millions of rural inhabitants have migrated to cities, providing an abundant, low-cost labor force. This migration has fueled China's economic boom, generating a significant demographic dividend. However, this model has relied heavily on scaling up investments and increasing construction resources, which has strained the country's limited natural resources and weakened the ecological environment. As a result, challenges such as environmental pollution, ecological degradation, and water shortages have emerged [96].

Despite these pressures, China's population density in built-up areas does not rank among the highest globally [97]. When considering total population relative to the administrative boundaries of cities, China's overall urban density appears low. Major cities like Beijing and Shanghai have an average population density of about 1000 people per square kilometer—relatively modest compared to global urban centers (see Figs. 3.5 and 3.6). The exception is Shenzhen, where the population density surged to 8793 people per square kilometer in 2021. In contrast, Suzhou has a population density of just 1552 people per square kilometer.

Statistical data from 2006 to 2016 reveal a significant shift in the population density of urban built-up areas in China's inland regions (Table 3.4) [99]. Areas with a population density exceeding 10,000 people/km^2 decreased from 41.7% to 19.2%, while areas with densities below 7500 people/km^2 increased from 12.2% to 35.4% by 2016. This shift means that, in 2016, 35.4% of built-up areas had a density under 7500 people/km^2, 45.4% ranged from 7500 to 10,000 people/km^2, and 19.2% had a density greater than 10,000 people/km^2. Notably, 80.8% of these areas now fall under 10,000 people/km^2. This trend highlights a clear redistribution of the population as cities expand.

In 2021, China's urbanization rate reached 64.72%, and the challenges facing both large cities and rural areas have become pressing concerns. Major cities like Beijing, Shanghai, Guangzhou, and Shenzhen are grappling with rising populations, skyrocketing housing prices, and worsening environmental issues such as smog and

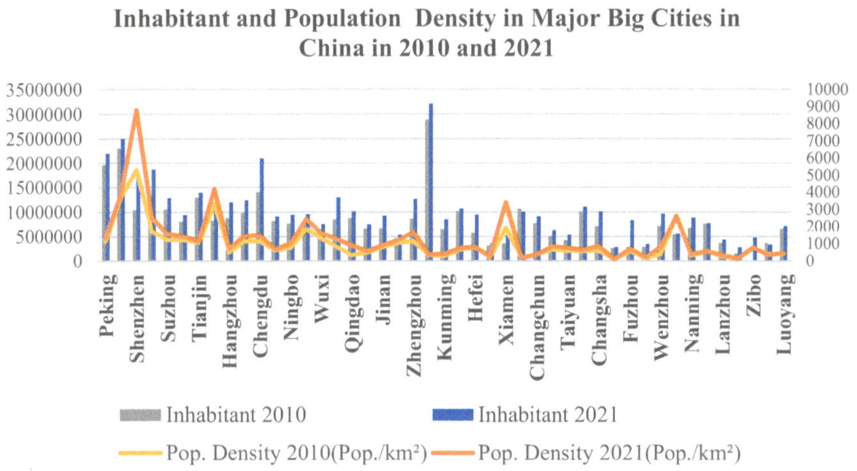

Fig. 3.5 Inhabitant and population density in major big cities in China in 2010 and 2021 (*Source* Diagram redraw by the data come from [97])

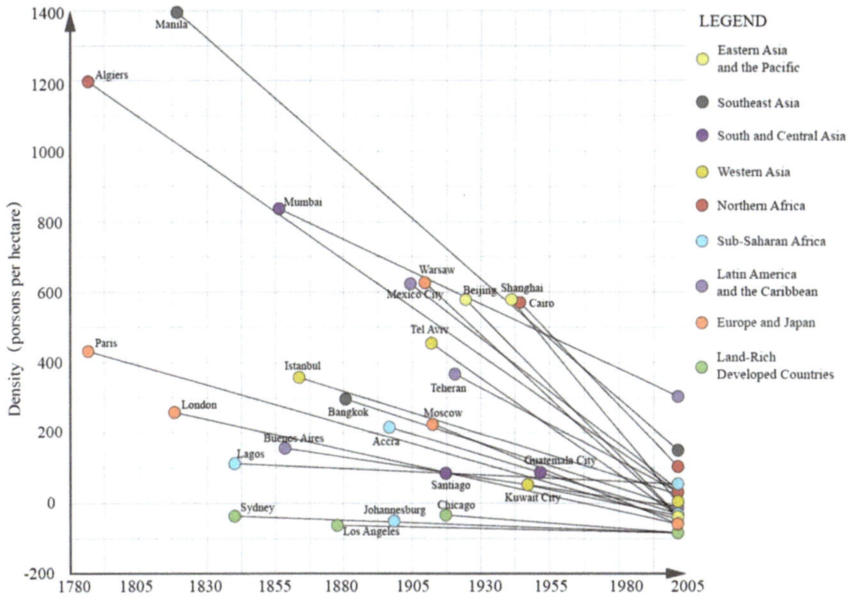

Fig. 3.6 The general decline in built-up area densities in 25 representative cities (*Source* Image adapted by the authors from [98])

Table 3.4 Population density changes in China from 2006 to 2016

Population density in urban built-up areas	2006 (%)	2016 (%)
More than 10,000 people/km^2	41.7	19.2
7500–10,000 people/km^2	12.2	35.4
Less than 7500 people/km^2	46.1	45.4

congestion. In contrast, rural areas in the central and western regions are stagnating, facing problems like the aging "empty nest" population, left-behind children, and the increasing vulnerability of migrant workers [100].

As urbanization progresses and undeveloped space becomes increasingly limited, the high-density urban model has emerged as a necessary solution for China's sustainable development. The demand for optimized land use is urgent (Table 3.5). Several scholars argue that high-density development offers several benefits:

Proximity fosters community interaction, reducing social isolation, particularly for vulnerable groups such as old people.
Compact settlements decrease the need for transportation, leading to reduced car usage, which benefits both public health and the environment.
Higher density reduces environmental impact, lowering carbon emissions.

In China, however, high-density urbanization must account not only for geographical constraints but also for the varying levels of attractiveness across different regions. Since the country's opening-up, eastern coastal cities have concentrated wealth and population, driven by central government policies. Geographer Huanyong Hu's "Huhuanyong Line" (Fig. 3.7) reveals a clear pattern in China's population distribution, with 96% of the population living to the east of the line. Over the past four decades, five core metropolitan areas have developed, with the eastern coast serving as the focal point, connected by vital transport routes such as the Yangtze River and Beijing-Guangzhou lines, forming a strategic economic bow.

Looking ahead, it is projected that nearly one billion people will reside to the east of the Huhuanyong Line, converging toward the central cities in the region known as the "Peacock Flying Southeast" [102]. This population concentration is expected to be concentrated within three major urban agglomerations: the Beijing-Tianjin-Hebei, the Yangtze River Delta, and the Pearl River Delta (Fig. 3.7). The rapid growth in these areas will extend to the Guangdong-Hong Kong-Macau Greater Bay Area and the Beijing-Tianjin-Hebei urban agglomeration.

Table 3.5 The redevelopment of three typical regeneration projects in Shenzhen

Village name	Site Area	GFA	FAR (before)	FAR (after)
Dachong Village (2014)	36.4 ha	280 ha	4.41	7.7
Hubei Village (2018)	40 ha	205 ha	3.4	11.6
Baishizhou Old Town (2019)	48 ha	348 ha	7.25	11.5

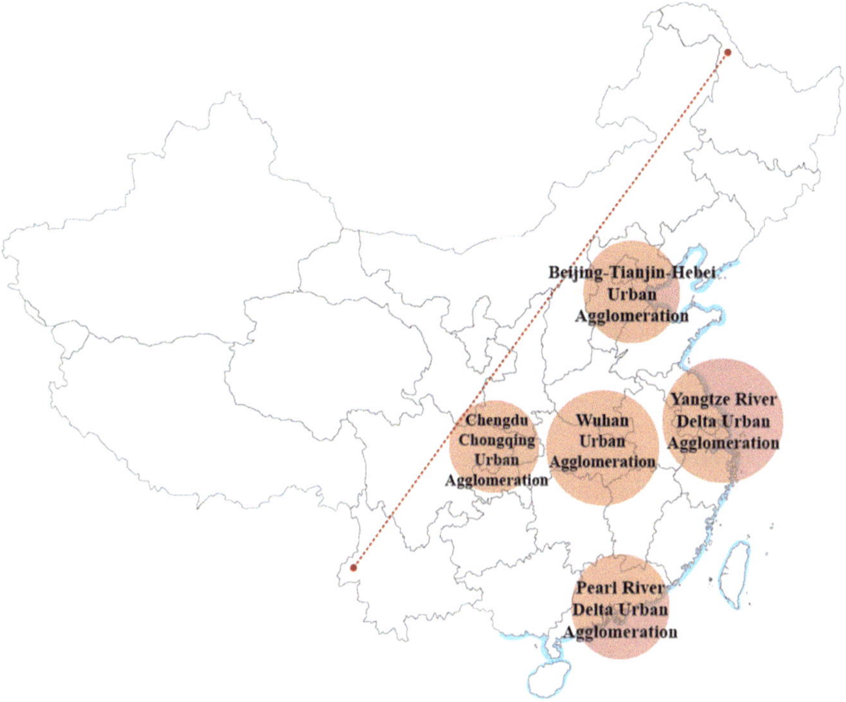

Fig. 3.7 "Huhuanyong Line" and urban agglomerations in China (*Source* Image adapted by the authors from [101])

To deal with the future increased concentration policies continue to evolve, reflecting ongoing research and development. [103] the newly issued National Territorial Spatial Plan clarifies the control of the "three lines and three areas" (三区三线), emphasizing the need to balance agricultural, ecological, and urban spaces while restricting construction land. The Plan sets three key boundaries at the regional level: the permanent basic farmland protection line, the urban development boundary, and the ecological protection line. This framework aims to strengthen ecological and environmental protection, with strict limitations on cultivated land and urban construction areas.

The agglomeration process in China is irreversible, a reality now increasingly accepted by scholars and the Chinese government. Further guidelines must support ongoing development and regeneration efforts. Consequently, scholars increasingly explore how to achieve sustainable urban development through moderate density in major urban centers [104]. At the same time, to meet financing requirements and maintain specific density levels, urban space quality must be enhanced.

3.5.3 "Moderate" Density: An Adapting Dimension to the Context Conditions

Densification serves as a powerful tool for urban regeneration, if it revitalizes existing areas and generates adaptive strategies for diverse contexts. However, determining the ideal density level and its distribution throughout a city remains a key debate in urban planning. Worldwide, urban population densities vary greatly, with no universally accepted benchmark for optimal density. Given the preceding discussions, it is clear that the appropriate level of densification in contemporary Chinese urban development must be defined contextually. Thus, this book propose the concept of moderate density to address the challenges of urban regeneration in China.

"Moderate" reflects a balance with the socio-economic context, steering clear of extremes when they do not align with local urban culture. The concept of moderation, which resonates deeply in Chinese public discourse, embodies the pursuit of harmony. Moderate density relies on a balance of development intensity, diversified open spaces, and multi-level urban interfaces, avoiding both hyper-density and sparse development. Scholars have already proposed this concept as a viable approach to urban development [105].

As China's urbanization progresses, the concept of moderate density must remain dynamic, continuously adapting to evolving circumstances.

To define an appropriate or "moderate" density level, several contemporary challenges facing urban China must be considered.

3.5.3.1 Development Limitation

China's population growth is nearing its peak, and the demographic inflection point. In addition to this clear and essential data, several related factors contribute to the complexity of defining urban density:

Urbanization Rate. With China's urbanization rate at 65%, achieving high-quality density is critical to ensuring residents' well-being.
Ageing Population: The aging demographic, coupled with declining purchasing power and consumption, has reduced housing demand.
Vacant Real Estate: Approximately 600 million square meters of newly constructed real estate remain unsold for over a year.
Limited Land for Redevelopment: The availability of land for large-scale urban transformation has dwindled, shifting the focus to the renewal of existing stock and urban land optimization [106].

Global urban trends reveal that ultra-high-density cities, such as Hong Kong and Shenzhen, face significant challenges in terms of urban safety, vitality, and energy consumption. These cities increasingly prioritize ecological sustainability, urban security, and vitality, advocating for medium-density development with a focus on open spaces, varied building heights, and balanced development intensity.

Therefore, it is necessary to carry out concentrated development and smart growth based on a reasonable balance between low carbon, low energy consumption, and high-density development. As existing literature clearly shows, concentration and combination of uses based on a density gradient, which allows traffic efficiency, and smart energy allocation and consumption must be the target. In addition, the integration of vertical mixing, micro green system, and public service facilities in the multi-functional plot promotes a pedestrian-friendly block and produces a vital center [107].

3.5.3.2 Ecological and Health Aspects

The epidemic has reshaped societal attitudes, political stances, and spatial dynamics, underscoring the role of compact cities in disease transmission [108]. While density is just one factor influencing vulnerability to the virus, it intersects with issues such as social stratification, mobility, social interaction, housing quality, infrastructure, and public spaces. High density fosters the convergence of urban elements and social ecosystems, but urban design must sustain diversity and heterogeneity to encourage endogenous urban evolution and balanced development.

In summary, both extreme high and low urban densities pose distinct challenges. There exists an inverted U-shaped relationship between urban density and economic, energy, and humanistic comfort. Theoretically, a moderate density can balance the disadvantages of both extremes, maximizing the benefits across economic, environmental, and social dimensions.

3.5.3.3 Avoid Extremes: Moderate Density

Urban density indicators, such as FAR, significantly influence economic benefits and transportation energy consumption. Optimal FAR values vary by city size, transportation mode, and climate, ranging from 1.5 to 3.0 (Fig. 3.8). Similarly, building height impacts cooling and heating energy consumption and urban safety, with optimal heights ranging from 18 to 45 m. Furthermore, building density and the open space ratio directly affect people's spatial comfort. Ensuring sufficient continuous building interfaces and increasing the open space ratio enhances comfort for residents.

The national government has expressed its concern about the current status of high-density cities, particularly regarding the well-being of their populations. Several official documents outline guidelines aimed at addressing the challenges related to density:

1. **Single buildings**: Restrict the approval of super high-rise buildings and support decision-making that favors moderately tall new developments.
2. **Residential communities in small cities**: Encourage the construction of residential buildings exceeding six floors in counties and smaller cities.

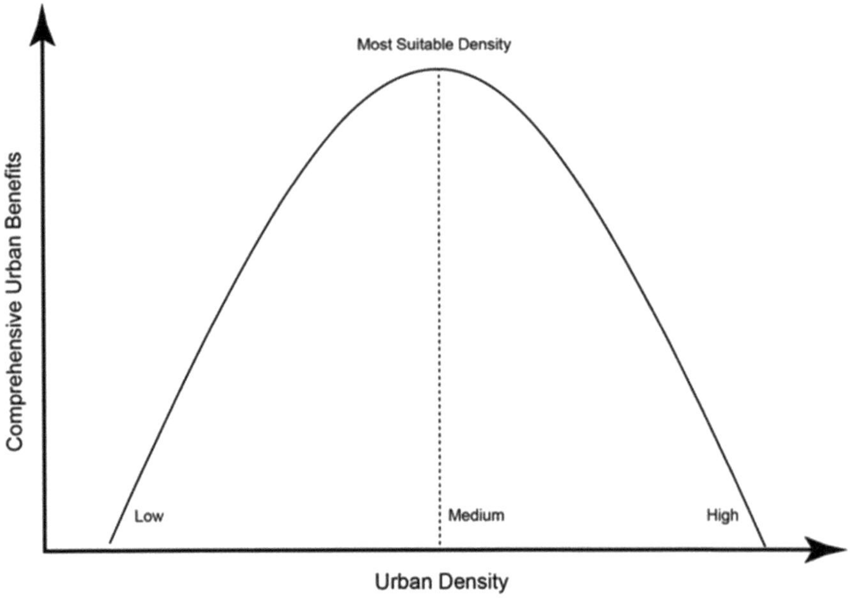

Fig. 3.8 Relationship between urban density and urban comprehensive benefits (*Source* Image adapted by the authors from [105, 109])

3. **Urban regeneration**: Implement strict controls over the built-to-area ratio and FAR across China's urban regeneration efforts [110].

These documents do not specify an ideal density level but highlight the importance of balance. As a benchmark for urban density, "medium density" can serve as a guiding metric, incorporating social, economic, environmental, and institutional considerations. Historically, China's large cities focused on alleviating housing and land shortages, resulting in higher medium-density values. However, with today's emphasis on ecological sustainability and people-oriented development, this medium density must be moderately reduced to address urban land and resource pressures. This shift not only helps conserve resources but also reduces energy consumption, health risks, and enhances urban livability.

Medium density control occurs at two levels: micro-level parcel density management and macro-level overall density zoning. The core variables for managing parcel density are plot ratio, building height, and open space ratio—the three key indicators examined in this study. Large cities can determine suitable medium-density guidelines based on their specific conditions. A medium development intensity of around 1.6–3.0 is recommended as a benchmark to regulate building heights (multi-story buildings under 36 m, medium and high-rise buildings up to 45 m, and high-rise buildings up to 60 m), as well as open space ratios (more than 60% for lower-density areas, and over 80% for higher-density areas) [111].

At the macro level, overall density zoning aims to achieve spatial differentiation and prevent excessive construction. This study proposes adopting "medium density" as an intermediate range to form an "olive-shaped" urban density zoning structure, where medium-density areas dominate, while high- and low-density areas remain limited [112, 113]. Establishing medium-density models at both city and plot levels can optimize economic, environmental, and social outcomes, providing valuable insights for future urban planning and density control.

In the context of China's rapid urbanization, it is crucial to recognize the gap between land and population urbanization. Land urbanization has outpaced population urbanization, necessitating a focus on controlling land expansion while optimizing the use of existing land resources. Cities must adopt strategies to curb the disorderly expansion of urban boundaries, enhance land use efficiency, and pave the way for more intensive, compact, and ecological urban development.

3.6 Conclusion

Drawing from the discussions on density, regeneration, and densification, several key conclusions emerge.

First, high density is not a problem but a fundamental aspect of urban development. As populations grow and land use restrictions increase, the intensity of land use in metropolitan areas will continue to rise.

Second, the concept of density is diverse and becomes more complex as urban environments evolve. The relationship between urban elements and social organizations can be understood through various density indicators.

Third, higher building density does not necessarily compromise a good spatial quality [114]. Often, density alone does not determine the quality of the urban form, but its relationship with public spaces and supporting infrastructure.

Fourth, "moderate" density must be context-driven. It requires a balance of development intensity, diverse open spaces, and multi-layered urban interfaces, avoiding both hyper-density and low-density extremes.

References

1. Chen J, Pellegrini P, Yang Z, Wang H (2023) Strategies for sustainable urban renewal: community-scale GIS-based analysis for densification decision making. Sustainability 15:7901
2. Pan L, Gan W, Chen J, Ren K (2023) An integrated model for constructing urban ecological networks and identifying the ecological protection priority: a case study of Wujiang District, Suzhou. Sustainability 15:4487

3. Ji M (2015) Urban housing security satisfaction in China: An HLM model and empirical study based on the 2012 Lian's public service quality of Chinese cities survey. E-Government 150:67–75
4. State Council of the People's Republic of China (2021) Outline of the 14th Five-Year Plan (2021–2025) for national economic and social development and long-range objectives through the year 2035 [R]. Central Party Literature Publishing House, Beijing. Translated by CCPC Party History and Literature Research Institute (Official English version)
5. Ma G, Paola P, Wu H, Han H, Wang D, Chen J (2023) Impact of land-use mixing on the vitality of urban parks: evidence from big data analysis in Suzhou, Yangtze River Delta Region, China. J Urban Plan Dev 149:4023045. https://doi.org/10.1061/JUPDDM.UPENG-4334
6. Chen J, Tian W, Xu K, Pellegrini P (2022) Testing small-scale vitality measurement based on 5D model assessment with multi-source data: a resettlement community case in Suzhou. ISPRS Int J Geo Inf 11:626. https://doi.org/10.3390/ijgi11120626
7. Chen J, Pellegrini P, Wang H (2022) Comparative residents' satisfaction evaluation for socially sustainable regeneration—the case of two high-density communities in Suzhou. Land 11:1483. https://doi.org/10.3390/land11091483
8. Groulx M, Nowak N, Levy K, Booth A (2021) Community needs and interests in university-community partnerships for sustainable development. Int J Sustain High Educ 22:274–290
9. Zhang X, Lu Y, Xu Y, Zhou C, Zou Y (2024) Governing regional inequality through regional cooperation? A case study of the Guangdong-Hong Kong-Macau Greater Bay Area. Appl Geogr 162:103135. https://doi.org/10.1016/j.apgeog.2023.103135
10. Zhang X, Zhou Z, Song Y (2024) Contested geopolitics of distribution in the city-region-building: promoting enclave industrial parks in the Pearl River Delta, China. Trans Inst Br Geogr 50(2):e12697
11. Churchman A (1999) Disentangling the concept of density. J Plan Lit 13:389–411
12. Tikoudis I, Farrow K, Mebiame RM, Oueslati W (2022) Beyond average population density: measuring sprawl with density-allocation indicators. Land Use Policy 112:105832
13. Zhang X, Sun Y, Luan X (2023) Regulation flexibility and legitimacy building in governing intercity railways: the polymorphous role of the Chinese provincial government. J Urban Aff 1–16
14. Ng E (2009) Designing high-density cities: for social and environmental sustainability. Routledge. ISBN 1136546006
15. Vale B, Vale R (2009) Is the high-density city the only option? In: Designing high-density cities for social and environmental sustainability. Routledge, pp 19–26
16. Clark G, Moonen T (2015, Octobre) Density: drivers, dividends and debates: appendix two: survey results
17. Zhang J (2012) The enlightenment of the western land development right system to the reform of my country's land management system. China Econ Trade Her 3
18. Zhu J (2012) From land use rights to land development rights: institutional change in China's urban development. In: The great urbanization of China. World Scientific, pp 191–220
19. Xu G, Jiao L, Yuan M, Dong T, Zhang B, Du C (2019) How does urban population density decline over time? An exponential model for Chinese cities with international comparisons. Landsc Urban Plan 183:59–67
20. Suzhou Municipal Bureau of Statistics (2021) Suzhou statistical yearbook 2021 [R]. China Statistics Press, Suzhou. ISBN 978-7-5037-9567-2
21. Wen L, Kenworthy J, Marinova D (2020) Higher density environments and the critical role of city streets as public open spaces. Sustainability 12:8896
22. Zhang X, Chen J, Wang H, Yang D (2025) From policy synergy to equitable greenspace: unveiling the multifaceted effects of regional cooperation upon urban greenspace exposure inequality in China's megacity-regions. Appl Geogr 174:103472. https://doi.org/10.1016/j.apgeog.2024.103472
23. Chen J, Li P, Wang H (2025) Assessment of influential factors on commute and life satisfaction in a historic campus-adjacent environment: evidence from a comparison study of twin cities. J Urban Plan Dev 151:4024065. https://doi.org/10.1061/JUPDDM.UPENG-5171

24. Harris RJ, Longley PA (2000) New data and approaches for urban analysis: modelling residential densities. Trans GIS 4:217–234
25. Sun B, Yin C, Yao X (2022) Densification and health in China: a U-shaped association between population density and obesity. Trans Plan Urban Res 27541223221111500
26. Chen J, Ren K, Li P, Wang H, Zhou P (2024) Toward effective urban regeneration post-COVID-19: urban vitality assessment to evaluate people preferences and place settings integrating LBSNs and POI. Environ Dev Sustain. https://doi.org/10.1007/s10668-024-05378-8
27. Zhao M, Liu N, Chen J, Wang D, Li P, Yang D, Zhou P (2024) Navigating post-COVID-19 social–spatial inequity: unravelling the nexus between community conditions, social perception, and spatial differentiation. Land 13
28. Li M, Ye C (2015) Threshold standard and global distribution of high-density cities. World Reg Stud 1:38–45
29. Song H, Chen J, Li P (2025) Decoding the cultural heritage tourism landscape and visitor crowding behavior from the multidimensional embodied perspective: insights from Chinese classical gardens. Tour Manag 110:105180. https://doi.org/10.1016/j.tourman.2025.105180
30. Chen J, Gan W, Liu N, Li P, Wang H, Zhao X, Yang D (2024) Community quality evaluation for socially sustainable regeneration: a study using multi-sourced geospatial data and AI-based image semantic segmentation. ISPRS Int J Geo Inf 13. https://doi.org/10.3390/ijgi13050167
31. Zhao X, Chen J, Li J, Wang H, Zhang X, Yu F (2025) Unraveling the renewal priority of urban heritage communities via macro-micro dimensional assessment—a case study of Nanjing City, China. Sustain Cities Soc 124:106317
32. Baran P (1964) On distributed communications networks. IEEE Trans Commun Syst 12:1–9
33. Bodó B, Brekke JK, Hoepman JH (2021) Decentralisation: a multidisciplinary perspective. Internet Policy Rev 10:1–21
34. Montavon M, Steemers K, Cheng V, Compagnon R (2006) "La Ville Radieuse" by Le Corbusier once again a case study. In: Proceedings of the PLEA 2006—23rd international conference on passive and low energy architecture, conference proceedings, p 6
35. Wirth L (2017) Urbanism as a way of life. In *The city: critical essays in human geography*. University of Chicago Press, vol 44, pp 25–48
36. Roskamm N (2011) Das Konstrukt Dichte Und Die „europäische Stadt". In Die Zukunft der Europäischen Stadt. Springer, pp 71–85
37. Chen J, Zhao X, Wang H, Yan J, Yang D, Xie K (2024) Portraying heritage corridor dynamics and cultivating conservation strategies based on environment spatial model: an integration of multi-source data and image semantic segmentation. Herit Sci 12:419. https://doi.org/10.1186/s40494-024-01497-7
38. Chen J, Li P, Wang H (2024) MSPA-based collaborative evaluation and planning governance of territory and natural environment in metropolitan agglomerations—a case study of EGIDDZ in Yangtze River Delta. J Resour Ecol 15:1263–1273. https://doi.org/10.5814/j.issn.1674-764x.2024.05.014
39. Saarinen E (1943) The city: its growth, its decay, its future. Reinhold Publishing Corp
40. Chen J, Li P, Lei Y, Zhang Y, Lai C, Chen B, Liu J, Schnabel MA (2025) Leveraging augmented reality for historic streetscape regeneration decision making: a big and small data approach with social media and stakeholder participation integration. Cities. https://doi.org/10.1016/j.cities.2025.106214
41. Jacobs J (1961) The death and life of great American cities. Random House, New York. ISBN 067974195X
42. Gehl J (1996) Livet Mellem Husene, Udeaktiviteter Og Udemiljøer; Arkitektens Forlag, København. ISBN 9788774072805
43. Chen J, Wang H, Yang Z, Li P, Ma G, Zhao X (2023) Comparative spatial vitality evaluation of traditional settlements based on SUF: taking Anren ancient town's urban design as an example. Sustainability 15:8178
44. Burton E (2000) The compact city: just or just compact? A preliminary analysis. Urban Stud 37:1969–2001

45. Ellis C (2002) The new urbanism: critiques and rebuttals. J Urban Des 7:261–291
46. Downs A (2005) Smart growth: why we discuss it more than we do it. J Am Plan Assoc 71:367–378
47. Chuanglin F, Weifeng Q (2007) Research progress and thinking of compact city and its measurement methods. Urban Plan Forum 4:65–73
48. Perez LP, Sales ALP, da Silveira JAR (2020) Climate change and the unsustainable urbanism in the municipality of João Pessoa, PB, Brazil. Sustain Debate 11:304–340
49. Mensah CA (2014) Destruction of urban green spaces: a problem beyond urbanization in Kumasi City (Ghana). Am J Environ Prot 3:1
50. McLoughlin B (2008) Urban consolidation and urban sprawl: a question of density. Urban Policy Res 148–156
51. Meyer JR, Kain JF, Wohl M (2013) The urban transportation problem. In: The urban transportation problem. Harvard University Press
52. Quan SJ, Economou A, Grasl T, Yang PPJ (2020) An exploration of the relationship between density and building energy performance. Urban Des Int 25:92–112
53. Newman PG, Kenworthy JR (1989) Cities and automobile dependence: an international sourcebook. Gower Publishing
54. Wegener M, Fürst F (2004) Land-use transport interaction: state of the art. Available SSRN 1434678, p 119
55. Zhilin L, Bo Q (2013) Urban form and low-carbon cities: research progress and planning strategies. Urban Plan Int 4–11
56. Yu Z, Ming Y (2017) Analysis of the influence of passenger flow on the development intensity of rail transit stations in the core urban area of Shanghai. Archit J 22–26
57. Ng E, Chen L, Wang Y, Yuan C (2012) A study on the cooling effects of greening in a high-density city: an experience from Hong Kong. Build Environ 47:256–271
58. Güneralp B, Zhou Y, Ürge Vorsatz D, Gupta M, Yu S, Patel PL, Fragkias M, Li X, Seto KC (2017) Global scenarios of urban density and its impacts on building energy use through 2050. Proc Natl Acad Sci 114:8945–8950
59. Maloutas T (2024) The role of vertical segregation in urban social processes. Nat Cities 1:185–193
60. Boyko CT, Cooper R (2011) Clarifying and re-conceptualising density. Prog Plan 76:1–61
61. United Nations Department of Economic and Social Affairs (2023) The sustainable development goals report 2023 [R]. United Nations Publications, pp 1–67. https://doi.org/10.18356/9789210024354. Accessed 30 June 2025
62. Dovey K, Pafka E (2014) The urban density assemblage: modelling multiple measures. Urban Des Int 19:66–76
63. Cohen MA, Gutman M (2007) Density: an overview essay. Built Environ 33:141–144
64. Holman N, Mace A, Paccoud A, Sundaresan J (2015) Coordinating density; working through conviction, suspicion and pragmatism. Prog Plan 101:1–38
65. Organisation for Economic Co-operation and Development (2018) Rethinking urban sprawl: moving towards sustainable cities (policy highlights). Eur Spat Res Policy 27(1):281–291
66. Tonkiss F (2017) Commodity or commons? Knowledge, inequality, and the city. In: In the post-urban world: emergent transformation of cities and regions in the innovative global economy. Routledge, pp 210–222
67. Williams K, Joynt JLR, Hopkins D (2010) Adapting to climate change in the compact city: the suburban challenge. Built Environ 36:105–115
68. Mayson SS, Charlton S (2015) Accommodation and tenuous livelihoods in Johannesburg's inner city: the 'rooms' and 'spaces' typologies. In: Proceedings of the urban forum, vol 26. Springer, pp 343–372
69. Carmona M, Gallent N, Sarkar R (2010) Space standards: the benefits. CABE London, p 18
70. Dennis K, Urry J (2010) After the car. Choice Rev Online 47:3234
71. Raynor K, Matthews T, Mayere S (2015) Defining the density debate in Brisbane: how urban consolidation is represented in the media. In: Proceedings of the State of Australian Cities Conference 2015: Refereed Proceedings; State of Australian Cities Research Network, pp 1–11

72. Raynor KE (2017) Defining the density debate: social representations of urban consolidation in Brisbane.
73. Jenks M, Burton E, Williams K (2010) Compact cities and sustainability: an introduction. In: The compact city: E&FN SPON, London, pp 11–12
74. Makgetla N (2020) Inequality in South Africa—an overview. Pretoria Trade Ind. Policy Strat 54
75. Global Commission on the Economy and Climate (2013) Better growth, better climate: the new climate economy report [R]. World Resources Institute, Washington, DC. https://newclimateeconomy.report/2014/. Accessed 30 June 2025
76. Cai Z, Tang Y (2021) Toward a sustainable city: a scoping review of eco-cities development and practices in China. Chinese Urban Plan Constr 179–199
77. Greyling S, Berrisford S (2017) Using planning tools to enable informal settlement upgrading: identifying future opportunities for. Upgrad Informal SettlS S Afr Pursuing a Partnership-Based Approach 363
78. Glaeser E, Sims H (2015) Contagion, crime and congestion: overcoming the downsides of density
79. Organisation for Economic Co-operation and Development Urbanisation as Opportunity (2014) OECD regional outlook 2014. World Bank Publications pp 171–182
80. Croucher KL, Wallace A, Duffy S (2012) The influence of land use mix, density and urban design on health: a critical literature review. Centre for Housing Policy. Alison Wallace, Centre for Housing Policy Steven Duffy, Centre for Reviews and Dissemination
81. Li Y, Schubert S, Kropp JP, Rybski D (2020) On the influence of density and morphology on the urban heat island intensity. Nat Commun 11:1–9
82. Zhang K, Batterman S (2013) Air pollution and health risks due to vehicle traffic. Sci Total Environ 450:307–316
83. Wekesa BW, Steyn GS, Otieno FAO (2011) A review of physical and socio-economic characteristics and intervention approaches of informal settlements. Habitat Int 35:238–245
84. Berghauser Pont M, Marcus L (2014) Innovations in measuring density: from area and location density to accessible and perceived density. Nord J Archit Res 2:11–30
85. Zhang Z, Wen Q, Liu F, Zhao X, Liu B, Xu J, Yi L, Hu S, Wang X, Zuo L (2016) Urban expansion in china and its effect on cultivated land before and after initiating "reform and open policy." Sci China Earth Sci 59:1930–1945
86. Wey W-M (2019) Constructing urban dynamic transportation planning strategies for improving quality of life and urban sustainability under emerging growth management principles. Sustain Cities Soc 44:275–290
87. Forsyth A (2003) Measuring density: working definitions for residential density and building intensity. Des Br 9:2–8
88. Bertaud A (2019) Spatial distribution of land prices and densities: models developed by economists. In: Order without design. The MIT Press, vol 5, p 2016
89. Nakano A (2015) Elementos demográficos sobre a densidade urbana da produção imobiliária [M]. Editora da Universidade de São Paulo, São Paulo. ISBN 978-85-314-1571-6
90. Huchzermeyer M (2007) Tenement city: the emergence of multi-storey districts through large-scale private landlordism in Nairobi. Int J Urban Reg Res 31:714–732
91. Yang X, Xiao C, Jie Z (2018) Economic development efficiency and urban density inflection point: an empirical investigation based on global 5km spatial grid data. *Shar Qual Proc 2018 China Urban Plan Annu Conf* 1–16
92. Kaur M, Hewage K, Sadiq R (2020) Investigating the impacts of urban densification on buried water infrastructure through DPSIR framework. J Clean Prod 259:120897
93. Shao Q, Zhang W, Cao X, Yang J, Yin J (2020) Threshold and moderating effects of land use on metro ridership in Shenzhen: implications for TOD planning. J Transp Geogr 89:102878
94. Teller J (2021) Regulating urban densification: what factors should be used? Build Cities 2
95. Wang X, Shi R, Zhou Y (2020) Dynamics of urban sprawl and sustainable development in China. Socioecon Plan Sci 70:100736

96. Tan M, Li X, Lu C, Luo W, Kong X, Ma S (2008) Urban population densities and their policy implications in China. Habitat Int 32:471–484
97. World Bank Population, Total for China
98. Angel S, Parent J, Civco DL (2010) The persistent decline in urban densities: global and historical evidence of 'sprawl.' Lincoln Inst. L. policy Cambridge, p 151
99. Du C, Jiao L, Xu G (2018) Spatio-temporal patterns and driving factors of urban population density change of prefecture-level cities in china during 2006–2016. Trop Geogr 791–798
100. He C, Ye J (2014) Lonely sunsets: impacts of rural-urban migration on the left-behind elderly in rural China. Popul Space Place 20:352–369
101. Ding J (2022). Hu Huanyong's contributions to Chinese population: geography in commemoration of the 80th anniversary of discovery of the "Hu Huanyong line." In Chinese social sciences and humanities studies. World Scientific, pp 121–136
102. Wang X, Meng X, Long Y (2022) Projecting 1 Km-grid population distributions from 2020 to 2100 globally under shared socioeconomic pathways. Sci Data 9:1–13
103. Xu B, Jiang Y (2007) Optimize allocation and improve land plot ratio. China's Constr Old Revolut Basic Area 1
104. Hao L (2015) Overall architecture measures of sustainable urban spatial development model based on smart city. Sci Technol Inf 2
105. Zheng D, Dong S, Lin C (2021) The necessity and control strategy of "medium density" in metropolis. Int Urban Plan 36:1–9
106. Dong L, Wang Y, Lin J, Zhu E (2020) The community renewal of shantytown transformation in old industrial cities: evidence from Tiexi worker village in Shenyang, China. Chin Geogr Sci 30:1022–1038
107. Lin Y, Shui W, Li Z, Huang S, Wu K, Sun X, Liang J (2021) Green space optimization for rural vitality: insights for planning and policy. Land Use Policy 108:105545
108. Hamidi S, Sabouri S, Ewing R (2020) Does density aggravate the COVID-19 pandemic? Early findings and lessons for planners. J Am Plan Assoc 86:495–509
109. Couture V (2012) Valuing the consumption benefits of urban density. Job Mark Pap 70.
110. Ministry of Housing and Urban-Rural Development Circular of the Ministry of Housing and Urban Rural Development on Preventing Large-Scale Demolition and Construction in the Implementation of Urban Regeneration.
111. Ministry of Housing and Urban-Rural Development of China (2018). Unified standard for reliability design of building structures [S]. China Architecture & Building Press (Standard Code: GB 50068-2018). http://www.mohurd.gov.cn/. Accessed 30 June 2025
112. Cheng V (2009) Understanding density and high density. In: Designing high-density cities. Routledge, pp 37–51
113. Oliveira V (2021) 21st century cities. In: Proceedings of the ISUF 2020 virtual conference proceedings, vol. 1. Universidade do Porto, Portugal
114. Pellegrini, P., & Chen, J. (2025). The regeneration path not taken further: An experiment in urban densification and state entrepreneurialism in a resettlement neighborhood in Suzhou, China. *Urban Studies* 00420980251365188. https://doi.org/10.1177/00420980251365188.

Chapter 4
Socially Sustainable Regeneration: An Overview and Conceptual Framework

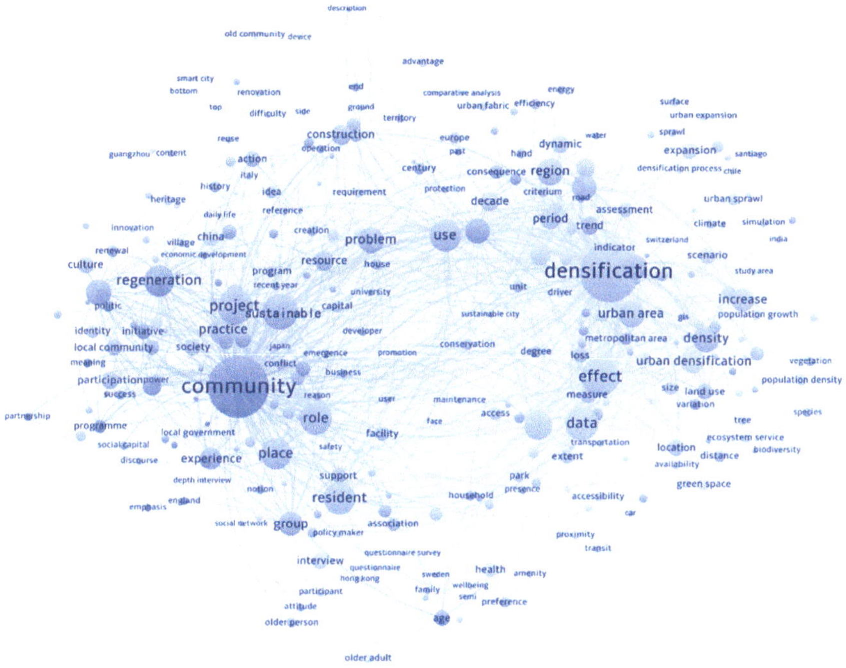

4.1 Introduction

The 2030 UN Agenda for Sustainable Development mentions 17 Sustainable Development Goals (SDGs); these goals aim to achieve global sustainable development in three dimensions: social, environmental, and economic [1]. These 17 SDGs correspond to challenges in poverty, climate, environment, etc., and the 11th goal focuses

J. Chen and P. Pellegrini, *Densification as a Design Tool for Sustainable Urban Regeneration*, Urban Sustainability, https://doi.org/10.1007/978-981-95-1653-7_4

on sustainable urban communities in response to the ongoing urbanization process. Through synthesizing the relevant sustainable urban regeneration policies and the dense exploration of regeneration contents, sustainability and regeneration-relevant study focused at the city level but lacked enough detailed research at the community level [2, 3].

The community as an integral part of sustainable development goals still lacks theoretical support, especially for the old community defined as built before 2000 in China [4, 5]. Specific explorations have addressed the problems in old communities, such as low public participation, inadequate infrastructure, and weakness of public space [6, 7]. The research involves so many disciplines and fields that a single theory cannot summaries it, including strengthening social participation in old communities [8], micro spatial regeneration [9], an organic transformation of public space [10], ageing-friendly [11], sustainable development, etc. However, as a microcosm of social problems, sustainable old community regeneration is an ongoing concern for all sectors of society [12]. There is an urgent need for a clear academic conceptual framework to guide relevant research [13].

This chapter proposes a theoretical framework by synthesizing the current research dynamics of old community regeneration, then analyzing and interpreting sustainable regeneration's research contents and methods from social, environmental, and economic perspectives [14]. Four intervention elements of old community regeneration have been exampled and discussed, such as land use, housing, infrastructure, and community culture. The discussion of social sustainability's vital role in community regeneration, and densification as an efficient and sustainable tool in contemporary theories and policies, provide a conceptual framework for guiding the empirical studies in the next stage [15].

4.2 Literature Overview Method

In order to explore the existing research on sustainable urban regeneration, the relevant works of literature were searched by the keywords such as *old community renewal*, *old community regeneration*, and *old community redevelopment.* These keywords have different meanings due to the other context. Still, the meanings expressed within them often have high similarities [16–18]. Therefore, in this chapter, the literature search about sustainable regeneration of old communities was not conducted with a single keyword, but with a combination of keywords: *old community, sustainable renewal, regeneration, and redevelopment.* The literature base data search was finalized in July 2022, and the data of the literature search from the year 2000 to 2022 mainly contained databases, namely Medline, Web of Science, SciELO Citation Index, Derwent Innovations Index, KCI-Korean Journal Database, and China Science Citation Database [19, 20].

As shown in Fig. 4.1, a total of 5002 documents were retrieved based on the search of database documents, including different research subjects worldwide. The retrieved documents were screened and sorted based on six aspects: abstract,

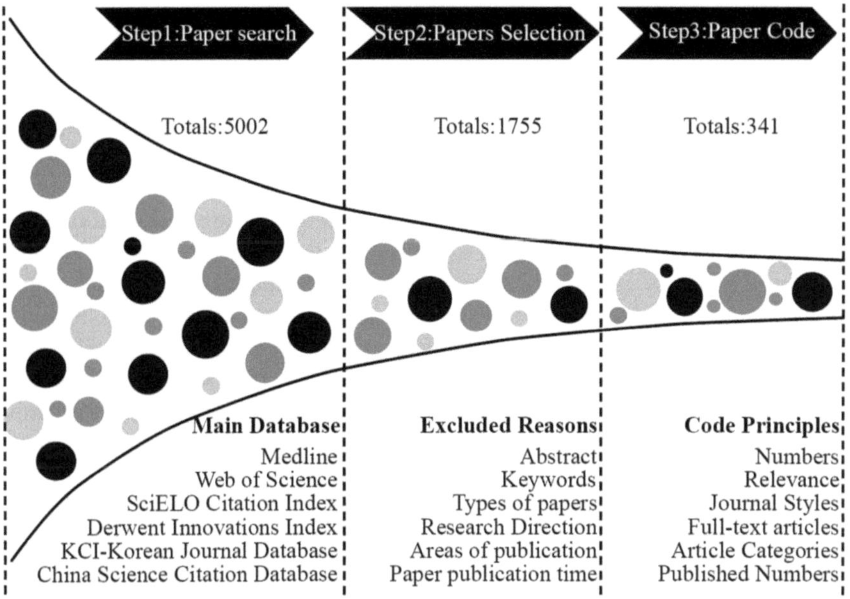

Fig. 4.1 Paper selection criteria and study process diagram

keywords, paper types, research direction, publication area, and publication time0. The judgement mainly relies on the papers' research direction related to keywords: *old community, sustainable renewal, regeneration, and redevelopment*. These papers were screened and categorized by keywords, *sustainability* and *old community regeneration*. The irrelevance samples in the abstract and keywords were removed for its not relevant to the research topic. A total of 1755 related papers were retained. After literature screening and coding according to literature content, journal category, and related content, a total of 341 highly relevant literature papers were obtained [21, 22].

4.3 Review of Current Studies

4.3.1 An Overview of Selected Literature

According to the 341 coded papers on the sustainable regeneration of old communities, it was found that the journals which published more papers on sustainable old community regeneration include *Sustainability, Land Use Policy, Journal of Korea Institute of Spatial Design, Biological Conservation Environment, Journal of the Architectural Institute of Korea Planning Design, Journal of the Urban Design Institute of Korea, Cities, Sustainable Cities and Society, Journal of Environmental*

Science and Management, etc. As shown in Table 4.1, the journals with more than four papers in this direction were selected for presentation. It mainly shows the distribution of the 341 papers in different journals.

The keyword network analysis of the code-completed 341 documents was conducted using VOSviewer software to generate a network analysis of sustainable old community regeneration through the relationship between keywords and the frequency of occurrence in different documents. In developing visualization using VOSviewer software, keywords with word frequencies higher than 40 were selected for display. It helps to understand the relationship between the keywords that recur in the articles, and the analysis identifies and captures the current trend of the research. Since there are relatively few articles from 2000–2012, the pattern is more evident in the figure to sort out the research lineage from 2012–2018. As shown in Fig. 4.2, the gradient from blue to yellow in the figure indicates the gradient of time from 2012 to 2018. After 2018, the literature numbers and contents are not that significant compared the previous period [23, 24].

From 2012 to 2015, the research focus shifted from renewal procedure and policy to community regeneration. From 2012, keywords such as social structure, policy, local government, and the process appeared more frequently in the relevant studies. Since 2015, the concepts of community, community localization, and regeneration have been gradually proposed, and the research scale has transitioned from city-based urban regeneration to community-based community regeneration. During this period,

Table 4.1 Overview of selected papers and the journals

Journal title	No	Journal title	No
Sustainability	79	The Science of the Total	7
Land Use Policy	16	Environmental	7
Journal of Korea Institute of Spatial Design	14	Urban Studies	7
Biological Conservation	13	Habitat International	7
Journal of the Architectural Institute of Korea Planning Design	12	Journal of the Architectural Institute of Korea	6
Journal of the Urban Design Institute of Korea	10	Journal of the Korea Institute of Culture Architecture	6
Cities	9	Ecology and Society	5
Sustainable Cities and Society	9	Journal of Green Building	5
Journal of Environmental Management	8	Journal of Sustainable Tourism	5
Journal of European Planning Studies International	7	Journal of Urban Planning and Development	5
Land	7	Sustainable Development	5
Journal of Sustainable Forestry	7	Building and Environment	5
Sustainable Development and World Ecology	7	Journal of Housing and the Built Environment	5
Science of the Total Environment	7	……	<5

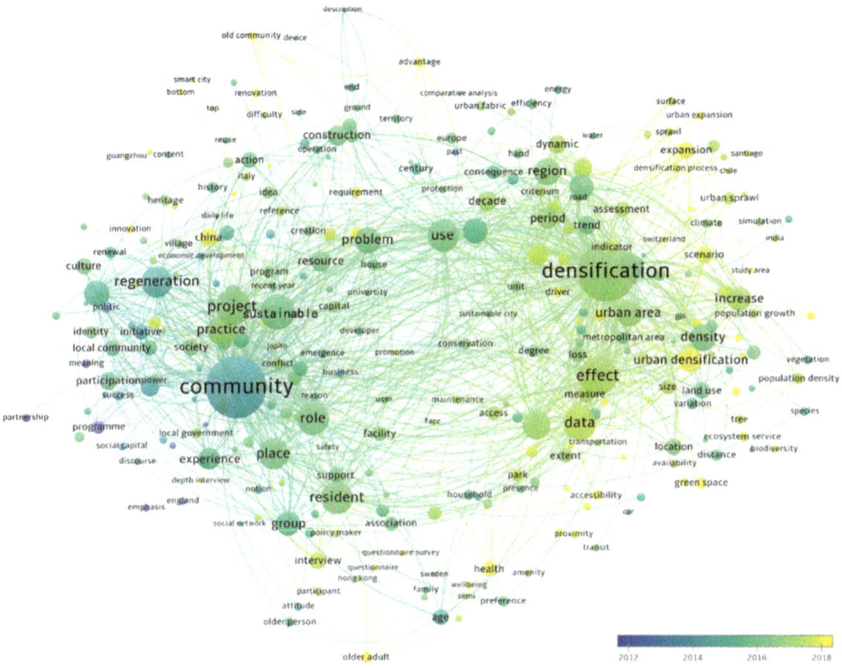

Fig. 4.2 Keywords co-occurrence mapping of research on old communities (*Source* Image by the authors)

the literature was also prominent for sustainable research, with related keywords such as sustainable, sustainable city, and the sustainable regeneration process. The essential contents emerged within the scope of sustainable regeneration, such as land use, housing, historical conservation, etc. Keywords about the high-density environment are also highlighted at this stage by many scholars. The regeneration process focused on relieving the pressure brought by the high-density built environments, such as density, population density, population growth, etc. [4, 7].

From 2016 to 2017, the discussion on density evolved, and keywords such as sprawl, surface, and densification were discussed incrementally. By 2018, the related discussion was gradually applied to the field of urban planning. In the urban regeneration process, as shown in the yellow part in Fig. 4.2, urban densification and expansion are frequently mentioned, among which the urban densification and densification process appear more regularly. At the same time, discussions about old communities, old adults, and renovation, such as health, requirement, green space, ecosystem service, access, accessibility, proximity, etc., are also frequently mentioned. The keywords indicate that the current community regeneration research tends to focus more on community participation, resident health, green space, and environmental diversity [5, 25, 26].

4.3.2 Three Aspects of Sustainable Community Regeneration Review

The old community regeneration has undergone the change from "renovation" to "regeneration," which not only reflects the current stage of old community regeneration focusing more on the needs of community residents for social participation and a green environment, but also reflects the different understanding of sustainability in the process of old community regeneration. As shown in Fig. 4.2, on this basis, this chapter compares the current research on sustainable old community regeneration through three different dimensions of sustainability: social, environmental, and economic.

A review of the current state of research, and geographical location both domestically and globally, reveals that the keyword *sustainable community regeneration* is the most frequently used in the process of old community regeneration. There are differences in the direction of sustainable community regeneration in different disciplines. Sustainable communities were initially considered to be the achievement of sustainable use of energy and resources[28]. As cities expand and population agglomeration proceeds, sustainable communities are not limited to resources[29]. The focus of sustainability gradually focuses on community health, environmental quality, and spatial justice[27]. As shown in Fig. 4.3, the research content can be divided into three categories: social sustainability, environmental sustainability, and economic sustainability, and the three influence each other and are interrelated [30, 31].

The socially sustainable related research focuses on quality of life, community development, equal opportunity, etc. As synthesized from the literature review, the socially sustainable at the community level much consider the relationship between individual needs and residents' satisfaction. The socially sustainable regeneration perspective study trend could be summarized as "Humanization and Precision" people-oriented regeneration. Environmentally sustainable research relates to resource management, environmental protection, habitat restoration, etc. The contemporary environmentally sustainable regeneration perspective research focuses on coordinating the overall community environment, which is "Complete Community" based regeneration. The economically sustainable research related to smart growth, long-range planning, investment saving and so on, at the community level focuses on creating a self-sustaining economic model: "Dynamically Equilibrium" financial support regeneration. The three dimensions of sustainability also involve somewhat different disciplines, with some overlap but not identical intercorrelation. As shown in Table 4.2, social sustainability focuses on community participation and community services. In contrast, environmental sustainability focuses on managing public space and environmental ecology, and economic sustainability focuses on studying community business models and market participation.

1. Social Sustainability: "Humanization and Precision" Old Community Regeneration

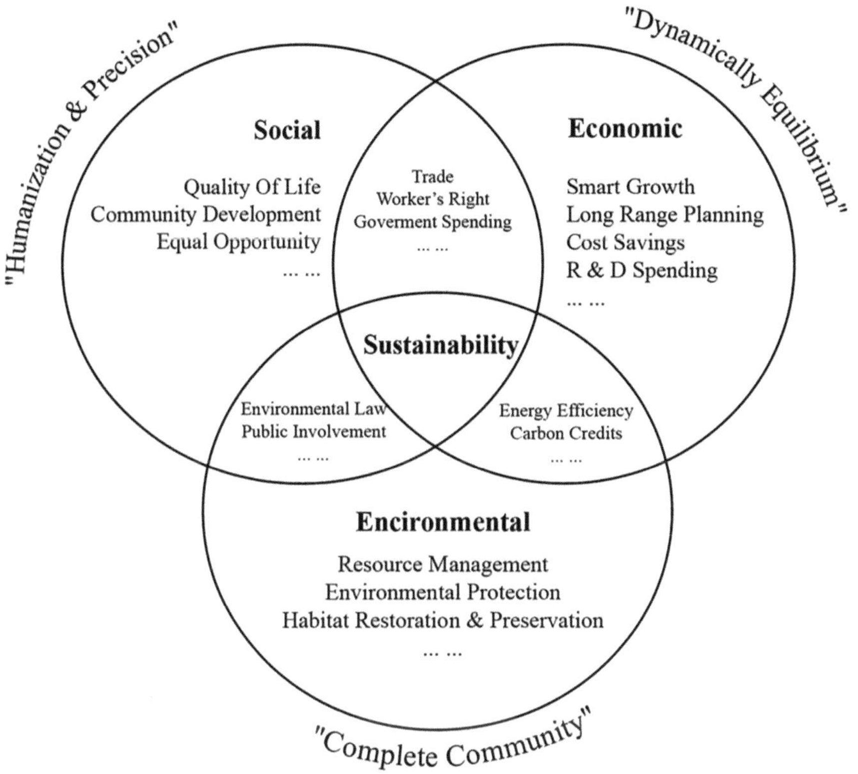

Fig. 4.3 Sustainability three aspects and relationships (*Source* Diagram redraw according to the [27])

Table 4.2 Major contents of densification categories at community level

Categories	Contents
Land use	Diversity functional land use, shopping, commerce, recreation and lofts
Housing	Building's quality and robustness, multiple functions, elevations adding, housing type
Infrastructure	Sponge and bioswale infrastructure, green infrastructure, aging friendly healing facilities, parking lots, accessibility
Culture	Public participation, neighborhood interactions, community management, residents' governance

The broader definition of social sustainability makes the connotation of cognition controversial. The initial concept of social sustainability was first proposed in 1987 by 'Our Common Future' [32], which emphasized sustainable development at the social level. Generally, socially sustainable communities can be defined as places where people want to live and work in the future. It contained three significant aspects: social equity, social capital, and social inclusion [33].

Regarding research content, the discussion on community social sustainability has increased in recent years. Social sustainability is the ability of communities to improve and renovate by community residents themselves, and it is an important strategy for urban regeneration [34]. Socially sustainable regeneration focuses on improving the quality of life, enhancing social participation and equity, and promoting community development [35]. Scholars pondered and interpreted the characteristics and connotations of socially sustainable development and believed that socially sustainable development should be on the construction of "quality of life" in urban communities [36]. And improve the community housing and built environment by obtaining residents' needs and strengthening residents' participation in community activities (Fig. 4.3).

Due to individual demands and differences in a social environment, factors affecting social sustainability in different old residential regeneration cases are significantly different [35]. Therefore, it is not suggested to formulate the old community regeneration strategy in a certain pattern. It is necessary to dynamically carry out the sustainable regeneration intervention of old communities from the spatiotemporal dimensions [37, 38]. To achieve social sustainability requires relevant mechanisms and strategies to improve residents' quality of life from the perspective of community integration and neighborhood interaction [39, 40], fully find out the community development potential through community participation [41]. In general, the current research on social sustainability in the urban planning field mainly focuses on the mechanisms and policies at the urban macro level, while fewer studies are concerned with the measurement system and factors influencing residents' perception at the micro-community level.

Multiple variables could assess social sustainability, such as social impact assessment (SIA), health impact assessment (HIA), environmental impact assessment (EIA), equity impact assessment (EqIA), sustainability appraisal (SA) and so on. Deakin proposed a people-centered and place-based assessment system [26]. Awad extracted the planning elements of urban regeneration through AHP (Analytic Hierarchy Process) [42]. The data analysis methods rely on the structural equation model (SEM) [43], regression model [30, 31], and other models to investigate residents' satisfaction [44] and well-being [45], etc. The SEM method overcomes the limitations of traditional statistical methods, making it an important tool for multivariate data analyses. The research methods tend to be humanistic and precise research with individual community residents as the unit.

In summary, the research tends to focus on the social sustainability of old community regeneration, continuously paying attention to the dynamic relationship between residents and the community, pinpointing the residents' needs, and strengthening public participation within the community.

2. Environmental Sustainability: "Complete Community" Old Community Regeneration

Environmental sustainability is increasingly recognized for its value in maintaining urban development quality. The geographical and environmental characteristics of the physical space influence sustainability [46]. At the community regeneration level, environmental sustainability refers to building multiple community facilities, creating a livable community living environment, and constructing an integrated environmental system, service system, and governance system.

By sorting out the research content, studies on environmental sustainability mainly focus on analyzing the relationship between people and the environment and creating public space. Relevant studies emphasize that creating environmentally sustainable communities requires strengthening the relationship between people and nature and the need to reflect the ecological character of the landscape in the regional context [47]. Some researchers and practitioners believe environmental sustainability counts the residents' participation, enhancing individual spatial creation [48].

From the research methods perspective, it was found that environmentally sustainable community regeneration prefers quantitative dynamic analysis. To achieve a sustainable built environment, natural elements such as green spaces and ecological infrastructures from microscales and mesoscale should be highlighted. Also, quantitative methods such as machine learning [49] and dynamic networks [50] provide a specific technical basis for the corresponding planning strategies, etc.

The "Complete Community Construction Standard" (CCCS) promulgated in January 2022 in China was in line with the study of environmental sustainability in old communities [51]. In the CCCS, the objectives of building a complete community include integrated public service facilities, sound commercial service facilities, complete municipal support facilities, sufficient public activity space, full-coverage property management, and a well-organized community management mechanism. With the continuous improvement of the "complete community" concept and the enrichment of its connotation, the concept further included the old community regeneration. Under the "complete community" concept, old community regeneration tends to develop into a systematic upgrading.

3. Economically Sustainable: "Dynamically Equilibrium" Old Community Regeneration

The economic sustainable development aims to maximize the quality of a comprehensive built environment with limited investment [36]. With the development of the digital economy, the bottom-up sustainable regeneration model of voluntary application by residents has gradually taken shape. Through multiple interests' cooperation, such as residents, market, finance, and policy, the four parties share the bottom-up model of sustainable regeneration. This mechanism has strengthened the public participation of community residents.

The issues of financing and benefit distribution have been the key issues in the economically sustainable old community regeneration. From the research content, the exploration of the economic model of community "Agriculturalization" [52], the implementation ability of action economy, the new economic attributes [53], and the study of the balance of community values [54], etc., aims to create a sustainable economic model through multiple perspectives to realize the housing value-added and spatial value-added in old communities [55].

The economically sustainable regeneration research methods are based on data analysis by establishing corresponding models, such as traditional economic growth models [56], stakeholder models [57], etc. Also, it could analyze the economic models of traditional communities using on-site interviews and follow-up surveys or by collecting community economic data through new devices, such as Web hotspots and Wi-Fi probes.

To sum up, sustainable old community regeneration requires a comprehensive analysis based on detailed surveys of all stakeholders, the elderly and children, the disabled, low-income, and other vulnerable groups. Sustainable regeneration needs to take full consideration of buildings, facilities, infrastructure, public facilities, and historical resources [58]. The relevant regeneration decision-making should extend the situ collective memory through a precise and comprehensive analysis of the diversity of residents. In conclusion, it needs to explore the diversified demands of the residents and the possibilities of spatial improvement, and also respect the wishes of communication, and provide multi-scenario space creation solutions:

Clarify the rules of equity distribution;
Promote the implementation of the theme of long-term localised operation;
continue to promote the optimisation of the business model.

In this way, it positively interacts with the residents, thus realising the economically sustainable old community regeneration.

4.3.3 "Social Sustainability": Community Regeneration Orientation

There is a wealth of research on the sustainable regeneration of old communities at different stages. Social, environmental, and economic sustainability are interlinked and influence each other. Studies tend to be integrated and coordinated, arguing that a more inclusive concept is needed to promote community regeneration. The environmental sustainability perspective focuses on analyzing the community environment, researching community public space and public facilities, and studies related to complete communities. The most prominent studies relate to complete communities' construction standards (CCCS). In contrast, the economic sustainability perspective tends to focus on creating profitable forms of the community economy. The research on the old community regeneration from a social sustainability perspective tends to study the relationship between individuals and communities, create a community culture, and optimize and upgrade old communities from a humanistic perspective [34].

The research methods on the old community regeneration at this stage mainly adopt mixed methods. The survey method shifted from traditional field surveys with questionnaires and photo interviews to multiple data-driven analyses using spatiotemporal positioning data, such as street view data, mobile phone signaling data, GPS positioning data and so on. A combination of "large" and "small" data [59], cross-validation analysis of residents' activities effectively enhances the cognitive identification of community space. It improves the objectivity and accuracy of spatial identification. For example, the old city regeneration in Nanjing, China, based on deep learning algorithms of artificial intelligence technology to evaluate aspects such as comfort, safety, and goodness. Through passive non-intervention collection of satellite positioning-based signaling dat to achieve a sense of community demographic structure and spatial distribution of people at different times [60].

Compared with environmental and economic sustainability, social sustainability is more concerned with improving quality of life, community development and equal opportunity. Also, social sustainability has been recognized as important for achieving high-quality urbanization and people-centered development since the Chinese National 13th Five-Year Plan (2015–2019) [61]. In terms of implementation difficulties, improving the physical environment on a community scale cannot achieve environmental and ecological sustainability. On the contrary, social sustainability is the community regeneration orientation, for its promoted at the community level through improving the quality of life, such as the physical environment improvement, infrastructure upgrading, including public space regeneration; and strengthening public participation, such like residents' satisfaction evaluation.

From the community intervention level, the social sustainability of old communities is more concerned with the residents' individual perception measurement of the conduct of the corresponding regeneration implementation. Engaging the public participation and residents' evaluation, with the precise regeneration needs, the implementation actions could be more suitable, which is less challenging to implement and more effective (Fig. 4.4). Social sustainability at the community level is fundamentally concerned with the stakeholders' perception; it can be strengthened by combining residents' satisfaction and professional planning & design disciplinaries. After this integrated operational framework, the relevant solutions and implementation strategies came out for improving the community's quality of life.

Fig. 4.4 The loop of the operational framework based on and for regeneration and social sustainability

Therefore, social sustainability is an integrated and inclusive concept that combines the relevant issues in old community regeneration from an environmental and economic perspective. It is an urgent research orientation for old community regeneration. The following section discusses the densification strategy for assisting sustainable community regeneration and how it combines with community elements, such as land use, infrastructure, buildings, and culture.

4.4 "Densification": Community Sustainable Regeneration Strategy

4.4.1 Densification as a Synergistic and Inclusive Sustainable Strategy

Densification is one of the main strategies to achieve sustainable development. It does not deny the benefits of rationalizing population density nor limit itself to creating densely built spaces. It aims to harmonize with the community environment by increasing the density of public space infrastructure, optimizing the overlay of different community needs, releasing "redundant" community public space, enhancing public participation within the community, and thus contributing to the shaping of community culture.

The original approach to densification attempted to achieve social justice; Ghafur argues that by increasing the density of living space within a community, the integration of different classes of residents is facilitated, and a rational distribution of social resources is achieved, leading to the sustainable development of the community [62]. This is one of the earliest studies and explorations of community densification in the process of community regeneration. More scholars extend the densification for sustainable community development; for example, high quality of life requires a dense control of the internal environment and greenery, and a moderate amount of dense space and greenery can help create a more sustainable center [63]. In addition, the idea is to design communities as "garden cities", creating a community garden culture by creating many Siècle-style buildings.

Densification has been explored globally, for example, in Bandung, Indonesia [64], where a comparison of the energy consumption of three different forms of urban development found that a densified urban center could contribute to urban sustainability or in Kazan City, Russia [65], where the densification of community housing was explored as an effective means of achieving sustainability in a dynamic urban development process.

Most densification research focuses on the urban scale, particularly on the role of densification in improving the timeliness of transport or reducing environmental losses and greenhouse gas emissions. However, less research has been conducted on how to achieve sustainable regeneration through densification design at the community scale. Therefore, in the following section, further research is needed on

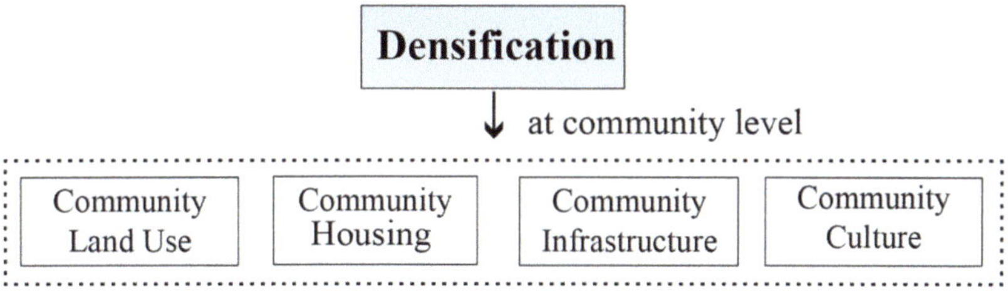

Fig. 4.5 Densification elements at community level

achieving intensive land regeneration through densification, strengthening the interaction between community residents and community spaces, and creating community cultural spaces with interactive attributes, thus forming densified and vibrant community living environment.

4.4.2 Densification Elements in Old Community Regeneration

In order to provide corresponding suggestions for the next stage of sustainability in old community regeneration, this section analyses and discuss four essential intervention elements of old community regeneration related to sustainability: land use, housing, infrastructure, and community culture (Fig. 4.5) [66].

4.4.2.1 Community Land Use

As integral to realizing sustainable community regeneration, Land Use Planning is not only a way of resource rearrangement but also a suitable and effective strategy [67]. It is found that by adapting and regenerating the stock land within the community area rather than demolishing and rebuilding, the contradiction between the community residents' spatial needs and the land use shortage can be effectively alleviated. In this way, community regeneration reduces the risk of ecological damage to the existing environment and saves resources [68].

Adaptive Land Use Planning can also provide some benefits to economic and environmental sustainability [69]. Land use conditions linked to the community residents' physical and mental health with the reasonable usage degree of land [70]. Monofunctional land use would reduce community exchange value and cause potential loss of housing value [71]. Mixed land use is necessary for realizing the sustainable form of cities. It represents the diversity of functional land use, such as housing, commerce, industry, etc. In the context of the community, we can understand it as increasing the functional diversity of the community and strengthening the connection between internal spatial elements [72]. Residents can meet the necessary daily needs in the community and form a self-sufficient community with less traffic behavior to reduce energy consumption [73].

4.4.2.2 Community Housing

Housing refers to residents' living spaces and determines daily life. Housing is a very sensitive indicator for sustainable regeneration, embodying most of internal everyday life [74]. The relevant policies and housing standards significantly affect the quality of life and positively or negatively impact sustainable community regeneration [74]. In addition, any change in housing will dramatically change the environmental and social structure of the community [74]. The density of community dwellings is a vital indicator for predicting, controlling, and assessing how people will use the space [75]. Under the rehabilitation and conservation background, the intensity of the current dwelling density could not easily achieve. The major transformation focuses on the architecture upgrading implementation; the treatment could have multiple possibilities, such as enhancing the building's robustness, injecting multiple functions, adding elevations and so on [76] (Fig. 2.9).

4.4.2.3 Community Infrastructure

Community infrastructure refers to the foundation of public realm quality, and infrastructure regeneration is one of the most critical strategies for fulfilling the needs of community residents [77]. The planning of community green infrastructure helps to promote the interactions between residents and the environment, and activate public participation in the community [78]. The different forms and functions of community infrastructure, such as sponge and bioswale infrastructure [79], and green infrastructure with vegetation [80], will significantly contribute to community engagement. The high-quality of community infrastructure will affect the essential health and well being of the residents with ageing-friendly healing facilities [81], it will also facilitate the achievement of equity in the community itself, including health equity, elimination of health disparities among them [82], etc.

4.4.2.4 Community Culture

Community culture is essential to improving cognition consensus, social cohesion, and public participation. Creating a community culture circle is conducive to evoking a sense of community identity and participation in various community affairs. Community culture is an important symbol and image of the community. The connotation of community culture could influence the regeneration orientation in the process of sustainable community regeneration. The community culture is created by the community stakeholders and reflects the residents' needs [83].

The participatory action research (PRA) in a low-income community in the UK proved that communities should enhance residents' sense of community involvement while improving quality of life and creating a community culture for achieving resilient communities [84]. By strengthening the community residents' public participation, infrastructure renovation's difficulty could be reduced, thus achieving sustainable community regeneration [85]. In summary, shaping community culture is conducive to strengthening social participation and achieving social sustainability in the community [86, 87].

4.5 Conceptual Framework for Sustainable Community Regeneration

As illustrated in the *Conceptual Framework diagram for Sustainable Community Regeneration,* in terms of the research contents, there are four parts of interpretation—fundamental research goal, basic components of a sustainable regeneration project, actions and tools, and relevant empirical studies at the community level (Fig. 4.6).

The first part, the general goal of the community regeneration research topic, starts from relevant discussions with the notions of sustainability and combines with the urgent new era of Chinese urbanization and local context, the regeneration movement in China and policies issued by the center and local government. Socially sustainable at the community level is the study goal and orientation in recent research. After confirming the purpose of community regeneration, the second part proposes the basic components of a sustainable regeneration project as the linkage of the research goal and research topic, which is the "output-transformation" interplay process. The major components of actions include four elements in the community: land use, housing, infrastructure, and culture (Table 4.2).

In the third part, Actions and Tools, this research "input" densification as a tool for promoting the regeneration process with its synergistic impact on the basic components in part two. On one side, as the urban regeneration policies and guidelines interpreted in Chap. 2, densification could solve the contradiction of limitation of land supply and high-quality living standard requirements. On the other hand, it refers to the synergistic and intensified of different forms and degrees of density through multiple processes; densification can result in a more vibrant and active mixed-use urban form. In terms of the specific densified objects could refer to Table 4.3.

Finally, in the empirical study part, an operational framework based on social sustainability is proposed (Figs. 4.4 and 4.6). The research expects to explore the relationship between design problems (results) and design strategies (inputs) through empirical studies. Based on the people-oriented development ideology and the integrated subjective and objective collaborative research methods, the research established satisfaction assessment and quantitative spatial analysis with stakeholders as

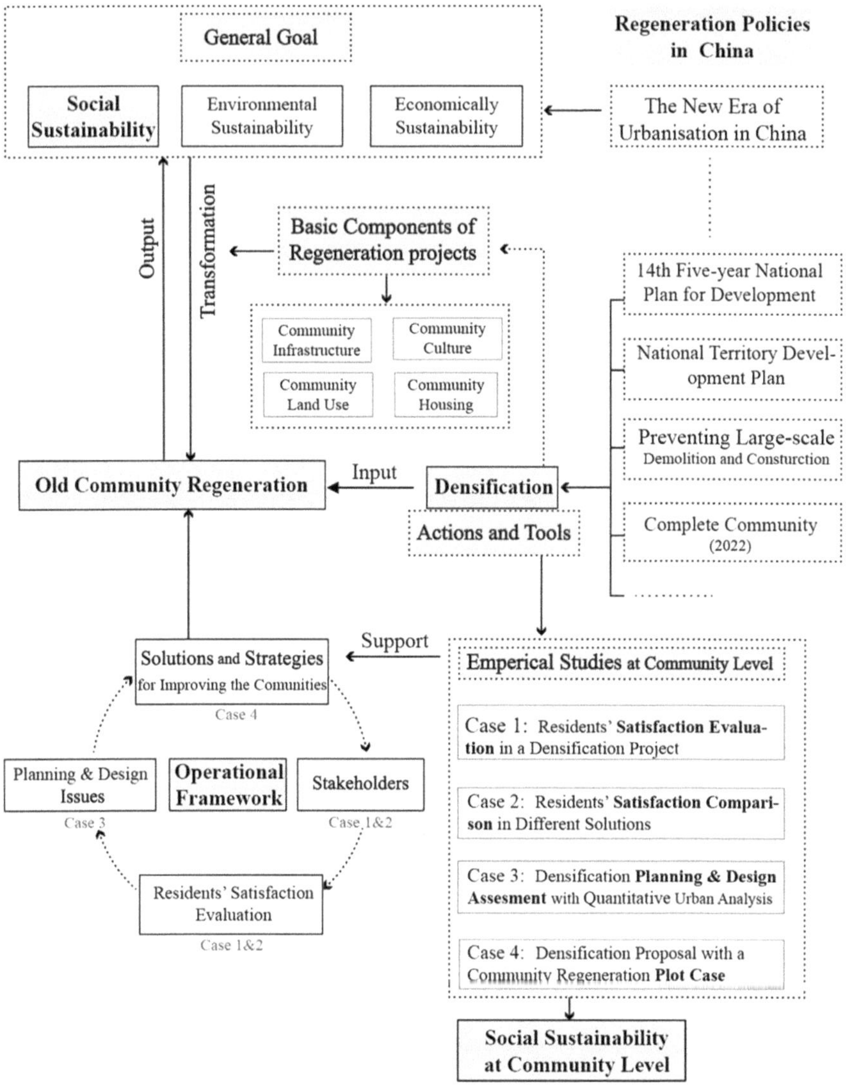

Fig. 4.6 Structure of conceptual framework diagram for sustainable community regeneration

the main body to identify the appropriate design solutions and strategies for sustainable community regeneration. Each case study could refer to one step in the operational framework, as shown in Fig. 4.6. This conceptual framework presents densification as a tool in promoting socially sustainable regeneration at the community level.

4.6 Conclusion

This study identified 341 papers on the sustainable regeneration of old communities and focuses on the research on old community regeneration in the context of sustainability from three perspectives: social, environmental, and economic. The relevant research review found that socially sustainable regeneration plays the most important role as it focuses on improving the residents' satisfaction and fulfilling living willingness. After reviewing all the content, this review indicates that further research is needed on achieving precision regeneration of community spaces. Humanistic and precision measurement of residents' satisfaction is a significant way to achieve sustainable community regeneration. Various new measurement methods emerging at present can effectively identify residents' perceptions more accurately and comprehensively.

This study also formed a conceptual framework diagram for sustainable community regeneration. Densification is a comprehensive and adaptive strategy for enhancing four key elements: land use, housing, infrastructure, and culture. The research methods in contemporary approaches to old community regeneration tend to identify and analyze community space by combining "large" and "small" data with mixed qualitative and quantitative analysis methods. With the people-oriented development ideology and the integrated subjective and objective collaborative research methods, the empirical studies should be conducted with a specific operational framework.

References

1. United Nations Department of Economic and Social Affairs (2023) The sustainable development goals report 2023 [R]. United Nations Publications, New York, pp 1–67. https://doi.org/10.18356/9789210024354. Accessed 30 June 2025
2. Zhang X, Zhou Z, Song Y (2024) Contested geopolitics of distribution in the city-region-building: promoting enclave industrial parks in the Pearl River Delta, China. Trans Inst Br Geogr 50(2):e12697
3. Zhang X, Sun Y, Luan X (2023) Regulation flexibility and legitimacy building in governing intercity railways: the polymorphous role of the Chinese provincial government. J Urban Aff 1–16
4. Chen J, Zhao X, Wang H, Yan J, Yang D, Xie K (2024) Portraying heritage corridor dynamics and cultivating conservation strategies based on environment spatial model: an integration of multi-source data and image semantic segmentation. Herit Sci 12:419. https://doi.org/10.1186/s40494-024-01497-7
5. Chen J, Ren K, Li P, Wang H, Zhou P (2024) Toward effective urban regeneration post-Covid-19: urban vitality assessment to evaluate people preferences and place settings integrating LBSNs and POI. Environ Dev Sustain. https://doi.org/10.1007/s10668-024-05378-8
6. Song H, Chen J, Li P (2025) Decoding the cultural heritage tourism landscape and visitor crowding behavior from the multidimensional embodied perspective: insights from Chinese classical gardens. Tour Manag 110:105180. https://doi.org/10.1016/j.tourman.2025.105180
7. Chen J, Gan W, Liu N, Li P, Wang H, Zhao X, Yang D (2024) Community quality evaluation for socially sustainable regeneration: a study using multi-sourced geospatial data and AI-based

image semantic segmentation. ISPRS Int J Geo-Inf 13(5):167. https://doi.org/10.3390/ijgi13050167
8. Hui ECM, Chen T, Lang W, Ou Y (2021) Urban community regeneration and community vitality revitalization through participatory planning in China. Cities 110:103072
9. Xi X, Xu H, Zhao Q, Zhao G (2021) Making rural micro-regeneration strategies based on resident perceptions and preferences for traditional village conservation and development: the case of Huangshan Village, China. Land 10:718
10. Fekete A, Hodor K, Dai D (2021) Urban sustainability through innovative open space design. A novel approach to the regeneration of historic open spaces in some Eastern European Countries and China. Earth 2:405–423. https://doi.org/10.3390/earth2030024
11. Jiang, C, Xiao, Y, Cao, H (2020) Co-creating for locality and sustainability: design-driven community regeneration strategy in Shanghai's old residential context. Sustain 12. https://doi.org/10.3390/su12072997
12. Zhao X, Chen J, Li J, Wang H, Zhang X, Yu F (2025) Unraveling the renewal priority of urban heritage communities via macro-micro dimensional assessment—a case study of Nanjing City, China. Sustain Cities Soc 124:106317
13. Shen T, Yao X, Wen F (2021) The urban regeneration engine model: an analytical framework and case study of the renewal of old communities. Land Use Policy 108:105571. https://doi.org/10.1016/j.landusepol.2021.105571
14. Zhang X, Chen Y, Xu Y, Yue W, Tang C (2023) Does regional cooperation constrain urban sprawl? Evidence from the Guangdong-Hong Kong-Macao Greater Bay Area. Landsc Urban Plan 235:104742
15. Groulx M, Nowak N, Levy K, Booth A (2021) Community needs and interests in university-community partnerships for sustainable development. Int J Sustain High Educ 22:274–290
16. Chen J, Pellegrini P, Yang Z, Wang H (2023) Strategies for sustainable urban renewal: community-scale GIS-based analysis for densification decision making. Sustainability 15:7901
17. Chen J, Li P, Lei Y, Zhang Y, Lai C, Chen B, Liu J, Schnabel MA (2025) Leveraging augmented reality for historic streetscape regeneration decision making: a big and small data approach with social media and stakeholder participation integration. Cities. https://doi.org/10.1016/j.cities.2025.106214
18. Ma G, Paola P, Wu H, Han H, Wang D, Chen J (2023) Impact of land-use mixing on the vitality of urban parks: evidence from big data analysis in Suzhou, Yangtze River Delta Region, China. J Urban Plan Dev 149:4023045. https://doi.org/10.1061/JUPDDM.UPENG-4334
19. Chen J, Tian W, Xu K, Pellegrini P (2022) Testing small-scale vitality measurement based on 5D model assessment with multi-source data: a resettlement community case in Suzhou. ISPRS Int J Geo Inf 11:626. https://doi.org/10.3390/ijgi11120626
20. Chen J, Pellegrini P, Wang H (2022) Comparative residents' satisfaction evaluation for socially sustainable regeneration the case of two high density communities in Suzhou. Land 11:1483 https://doi.org/10.3390/land11091483
21. Yang N, Chen J, Ban L, Li P, Wang H (2024) Pre-planning and post-evaluation approaches to sustainable vernacular architectural practice: a research-by-design study to building renovation in Shangri-La's Shanpian House, China. Sustainability 16:9568
22. Chen J, Wang H, Yang Z, Li P, Ma G, Zhao X (2023) Comparative spatial vitality evaluation of traditional settlements based on SUF: taking Anren ancient town's urban design as an example. Sustainability 15:8178
23. Chen J, Li P, Wang H (2024) MSPA-based collaborative evaluation and planning governance of territory and natural environment in metropolitan agglomerations—a case study of EGIDDZ in Yangtze River Delta. J Resour Ecol 15:1263–1273. https://doi.org/10.5814/j.issn.1674-764x.2024.05.014
24. Chenjing F, Yueying G, Chenxi J, Zhihao Y, Runhan L (2025) Does urban renewal enhance urban vitality? A case study of Nanjing, China, using multidimensional data. Environ Plan B Urban Anal City Sci 23998083251347680. https://doi.org/10.1177/23998083251347682
25. Hu J, Chen J, Li P, Yan J, Wang H (2024) Systematic review of socially sustainable and community regeneration: research traits, focal points, and future trajectories. Buildings 14:881. https://doi.org/10.3390/buildings14040881

26. Zhao M, Liu N, Chen J, Wang D, Li P, Yang D, Zhou P (2024) Navigating post-COVID-19 social–spatial inequity: unravelling the nexus between community conditions, social perception, and spatial differentiation. Land 13(4):563
27. Agyeman J (2005) Sustainable communities and the challenge of environmental justice. In: Sustainable communities and the challenge of environmental justice. New York University Press
28. Visa, I, Duta, A, Moldovan, M, Burduhos, B, Neagoe, M (2020) Sustainable communities. In: Green energy and technology, pp 341–384
29. Feenstra GW (1997) Local food systems and sustainable communities. Am J Altern Agric 12:28–36
30. Zhang X, Chen J, Wang H, Yang D (2025) From policy synergy to equitable greenspace: unveiling the multifaceted effects of regional cooperation upon urban greenspace exposure inequality in China's megacity-regions. Appl Geogr 174:103472. https://doi.org/10.1016/j.apgeog.2024.103472
31. Chen J, Li P, Wang H (2025) Assessment of influential factors on commute and life satisfaction in a historic campus-adjacent environment: evidence from a comparison study of twin cities. J Urban Plan Dev 151:4024065. https://doi.org/10.1061/JUPDDM.UPENG-5171
32. World Commission on Environment and Development (1987) Our common future. Oxford University Press
33. Ghahramanpouri A, Lamit H, Sedaghatnia S (2013) Urban social sustainability trends in research literature. Asian Soc Sci 9:185–193
34. Magee L, Scerri A, James P (2012) Measuring social sustainability: a community-centred approach. Appl Res Qual Life 7:239–261
35. Gu Z, Zhang X (2021) Framing social sustainability and justice claims in urban regeneration: a comparative analysis of two cases in Guangzhou. Land Use Policy 102:105224
36. Wanamaker C. The environmental, economic, and social components of sustainability
37. Liu J, Zhang Y, Ran A (2020) Research on the renewal and transformation of old districts in Beijing: based on characteristics-dilemma-policy analysis framework. Soc Gov Rev 10
38. Chen Y, Xiao L, Chen M, Jiang Y (2020) Preliminary exploration of approaches to community participatory planning: case study of 'New Qinghe Experiment' in Beijing. Urban Plan Forum 6
39. Zhang S, Lin S, Li Z, Guo Y (2019) The impact of neighborhood environment perception on self rated health of Chinese urban residents: a case study of Wuhan City. Hum Geogr 34:38–46
40. Li X, Li Z (2019) State-led social integration mechanism within new migrant enclaves of Chinese cities: a case of Dongguan. City Plan Rev 43:7
41. Glasson J, Wood G (2009) Urban regeneration and impact assessment for social sustainability. Impact Assess Proj Apprais 27:283–290. https://doi.org/10.3152/146155109X480358
42. Awad J, Jung C (2022) Extracting the planning elements for sustainable urban regeneration in Dubai with AHP (Analytic Hierarchy Process). Sustain Cities Soc 76:103496. https://doi.org/10.1016/j.scs.2021.103496
43. Tan Y, He J, Han H, Zhang W (2019) Evaluating residents' satisfaction with market-oriented urban village transformation: a case study of Yangji Village in Guangzhou, China. Cities 95:102394. https://doi.org/10.1016/j.cities.2019.102394
44. Ji X, Shao L, Du Y (2020) Collaborating with local communities to identify improvement priorities for historic urban landscape based on residents' satisfaction: an application of asymmetric impact-performance analysis in. Sustain Artic. https://doi.org/10.3390/su12041463
45. Gür M, Murat D, Sezer FŞ (2020) The effect of housing and neighborhood satisfaction on perception of happiness in Bursa, Turkey. J Hous Built Environ 35:679–697. https://doi.org/10.1007/s10901-019-09708-5
46. Dale A, Ling C, Newman L (2008) Does place matter? Sustainable community development in three Canadian communities. Ethics, Place Environ. 11:267–281. https://doi.org/10.1080/13668790802559676
47. Zhang X, Mao M, Xing X, Zhang P, Jiang D (2018) 'Thick data + big data' activates public life in old communities—a case study of Yaziqiao community in Beijing. Beijing Plan Rev 9

48. Hao Y, Wang Y, Wu Q, Sun S, Wang W, Cui M (2020) What affects residents' participation in the circular economy for sustainable development? Evidence from China. Sustain Dev 28:1251–1268
49. Wei L, Tian W, Silva EA, Choudhary R, Meng Q, Yang S (2015) Comparative study on machine learning for urban building energy analysis. Proc Eng 121:285–292
50. Goldbeck N, Angeloudis P, Ochieng WY (2019) Resilience assessment for interdependent urban infrastructure systems using dynamic network flow models. Reliab Eng Syst Saf 188:62–79
51. Ministry of Housing and Urban-Rural Development (2021) Complete residential community construction guide [R]. Ministry of Housing and Urban-Rural Development, Beijing (Document No.: *Jian Ban Ke* [2021] No. 55). https://www.mohurd.gov.cn. Accessed 30 June 2025
52. Liu C, Zhang Y, Zhao J (2018) Research on urban agro-integrated community based on circular metabolism. City Plan Rev 8
53. Liu J, Zhao Y (2021) Urban renewal: reconstruction or self-renewal?: A case study of Xiamen lakeside tansformation. Urban Dev Stud 28:1–6
54. Callaghan EG, Colton (ÆJ) Building sustainable & resilient communities: a balancing of community capital. 931–942. https://doi.org/10.1007/s10668-007-9093-4.
55. Lin C, Du Y, Yue J, Wang J (2019) A research on the urban regeneration mechanism of multiple-stakeholder collaboration: a case study of Shenzhen. Urban Plan Forum 06:56–62
56. Doll JP (1988) Traditional economic models of fishing vessels: a review with discussion. Mar Resour Econ 5:99–123
57. Clarkson ME (1995) A stakeholder framework for analyzing and evaluating corporate social performance. Acad Manag Rev 20:92–117
58. Chan E, Lee GKL (2008) Critical factors for improving social sustainability of urban renewal projects. Soc Indic Res 85:243–256. https://doi.org/10.1007/s11205-007-9089-3.
59. Martí P, García Mayor C, Serrano Estrada L (2019) Identifying opportunity places for urban regeneration through LBSNs. Cities 90:191–206
60. Kong Q (2019) Research on the aging adaptation design of old communities under the background of urban renewal—a case of Tianjin Xincun community in Nanjing. Urban Archit 16:2
61. State Council of the People's Republic of China (2023, June 12) Several opinions on deepening the construction of new urbanization [R] (Policy Document No.: Guo Fa [2023], No. 21). State Council General Office, Beijing. http://www.gov.cn/zhengce/content/2023-06/12/content_5746789.htm. Accessed 30 June 2025
62. Ghafur S (2006) Social exclusion and residential densification: implications for integration of the urban poor in Dhaka. Bangladesh 10:27–29
63. Tillie N, van Beurden JB, Doepel D Aarts M (2018) Exploring a stakeholder based urban densification and greening agenda for Rotterdam inner city—accelerating the transition to a liveable low carbon city. Sustainability.
64. Environment B, Permana AS, Aziz NA, Ho CS (2015) Three sustainability advantages of urban densification in a concentric urban form: evidence from Bandung City. Indonesia 2:158–167
65. Fedorova VA, Safina GR, Benedict EQ (2020) Environmental aspects of densification of residential development in urbanized territories (case study: Kazan City). Casp J Environ Sci 18:519–524
66. Lee Y, Lim S, Kim G (2010) Improvement characteristics shown in holistic regeneration of Ballymun toward sustainable community. Build Environ 45:279–286. https://doi.org/10.1016/j.buildenv.2009.08.013
67. Souza L, Bueno C (2022) City information modelling as a support decision tool for planning and management of cities: a systematic literature review and bibliometric analysis. Build Environ 207:108403. https://doi.org/10.1016/j.buildenv.2021.108403
68. Aytur SA, Rodriguez DA, Evenson KR, Catellier DJ, Rosamond WD (2007) Promoting active community environments through land use and transportation planning. Am J Heal Promot 21:397–407

69. Vargas Hernández JG, Orozco Quijano EP (2022) Sustainable resilience in urban land use. In: Handbook of research on sustainable development goals, climate change, and digitalization. IGI Glob 233–256
70. Wahidah AN, Martono DN (2022) Supriatna land use sustainability to mitigate potential land slide in Ciletuh Watershed, Sukabumi, Indonesia. IOP Conf Ser Earth Environ Sci 950. https://doi.org/10.1088/1755-1315/950/1/012006
71. Takahashi D, Fujie T, Senda T (2022) Conditions for collective land use by community-based organizations: case study of community farming enterprises in Japan. Int J Commons 16:209–224. https://doi.org/10.5334/ijc.1151
72. Chen J, Pellegrini P, Wang H, Ma G (2022) Sustainable urban form based old community regeneration—taking Jing'anli community in Tianjin as an example 2 . Transformation of old residential. Int J Archit Arts Appl 8:72–81. https://doi.org/10.11648/j.ijaaa.20220802.14
73. Alberti M (2000) Urban form and ecosystem dynamics: empirical evidence and practical implications. Achiev Sustain Urban Form 84–96
74. Winston N (2010) Regeneration for sustainable communities? Barriers to implementing sustainable housing in urban areas. Sustain Dev 18:319–330
75. Boyko CT, Cooper R (2012) High dwelling density as sustainability solution in Lancaster. Proc Inst Civ Eng Eng Sustain 165:81–88. https://doi.org/10.1680/ensu.2012.165.1.81
76. Matsumura S, Yue F, Tongtong L (2008) Reborn European American collective housing. China Machine Press
77. Brown J, Barber A (2012) Social infrastructure and sustainable urban communities. Proc Inst Civ Eng Eng Sustain 165:99–110
78. Mell IC (2009) Can green infrastructure promote urban sustainability? In: Proceedings of the Institution of Civil Engineers Engineering Sustainability (vol. 162). Thomas Telford Ltd, pp 23–34
79. Makropoulos CK, Butler D (2010) Distributed water infrastructure for sustainable communities. Water Resour Manag 24:2795–2816. https://doi.org/10.1007/s11269-010-9580-5
80. Baldauf R, McPherson G, Wheaton L, Zhang M, Cahill T, Bailey C, Fuller CH, Withycombe E, Titus K (2013) Integrating vegetation and green infrastructure into sustainable transportation planning. TR News 14–18
81. Williams P, Pocock B (2010) Building, "community" for different stages of life: physical and social infrastructure in master planned communities. Community Work Fam 13:71–87. https://doi.org/10.1080/13668800902903300
82. Stewart EC, Erves JC, Hargreaves MK, Duke JM, Sanderson M, Rowan N, Miller ST (2020) Developing an infrastructure to cultivate equitable and sustainable community-academic research partnerships: Meharry community engagement core. J Natl Med Assoc 112:423–427. https://doi.org/10.1016/j.jnma.2020.04.011
83. Roseland M (2000) Sustainable community development: integrating environmental, economic, and social objectives. Prog Plan 54:73–132. https://doi.org/10.1016/S0305-9006(00)00003-9
84. Cinderby S, Haq G, Cambridge H, Lock K (2016) Building community resilience: can everyone enjoy a good life? Local Environ 21:1252–1270. https://doi.org/10.1080/13549839.2015.1100597
85. Hyland M, Bertsch V (2018) The role of community involvement mechanisms in reducing resistance to energy infrastructure development. Ecol Econ 146:447–474. https://doi.org/10.1016/j.ecolecon.2017.11.016
86. Trickett EJ (2009) Multilevel community-based culturally situated interventions and community impact: an ecological perspective. Am J Community Psychol 43:257–266. https://doi.org/10.1007/s10464-009-9227-y
87. Marshall D, McCarthy L, McGrath P, Claudy M (2015) Going above and beyond: how sustainability culture and entrepreneurial orientation drive social sustainability supply chain practice adoption. Supply Chain Manag 20:434–454

Chapter 5
The Place of Innovations: Shenzhen Densification

5.1 Shenzhen Development and Shenzhen Speed

Shenzhen's urban history is remarkable. It has long served as a testing ground for urban solutions in China, with planners and designers recognizing its potential to shape future urban development models. As such, Shenzhen plays a central role in this research. Densification has already occurred here—and continues to unfold [1–3].

J. Chen and P. Pellegrini, *Densification as a Design Tool for Sustainable Urban Regeneration*, Urban Sustainability, https://doi.org/10.1007/978-981-95-1653-7_5

From a small fishing village to a sprawling metropolis in just 40 years, Shenzhen's development is unparalleled [2, 4]. Today, over 20 million people call Shenzhen home, drawn by its rapid urbanization and densification. In 2020, Shenzhen ranked among the world's top 10 in sustainable city competitiveness, claiming the top spot among all Chinese cities [5]. This highlights a dual achievement: rapid growth alongside a commitment to sustainable development. Thus, Shenzhen serves as an ideal case study for exploring sustainable densification.

When considering housing in Shenzhen, the focus must be on urban villages, which house over 13 million residents [6]. A decade ago, city managers viewed these villages as a "cancer" in urban development—associated with "dirt, disorder, and poverty" and problems like gambling and crime [7, 8]. Consequently, local government efforts concentrated on remediating illegal businesses, unlicensed shops, and unauthorized structures [9, 10]. In an attempt to improve the villages' image, government departments implemented cosmetic urban renewal programs, such as the "clothing and hat" building and landscape retrofit [9, 11].

However, such superficial fixes cannot address the root problems. Urban villages, located in prime city areas, often have higher plot ratios than traditional residential neighborhoods [12]. While the government has invested heavily in infrastructure and public services in and around these areas, it struggles to capture the land value generated by the private ownership structures of urban villages.

The widespread existence of urban villages in Shenzhen has both historical and practical causes [13]. Government regulations and policies have failed to keep pace with the explosive growth of these areas [8]. Rapid economic development and population influx have outpaced land transfer processes and the construction of commercial housing, leading to a severe housing shortage. This imbalance has created significant practical contradictions [14].

The ambiguity of policies and the diversification of policy understanding in some specific periods have led to the crazy expansion of the "rush to build and protect the land" [15, 16]. Substantial business opportunities and profit margins have led to a rapid influx of private funds [2, 17].

5.2 Regeneration and Densification with Demolition and Reconstruction: Baishizhou

Creating a sense of place in urban spaces requires a gradual accumulation of various factors, which serve as essential prerequisites for spatial transformation [4, 16]. These prerequisites include current functional density, spatial accumulation, architectural diversity, and mixed land use. In the context of rapid urbanization, densification through "demolition and reconstruction" serves as a critical method for revitalizing urban areas and stimulating regional vitality (Fig. 5.1).

Phase 1: Initially, the area consists primarily of old villages, with some self-built commercial and basic convenience structures.

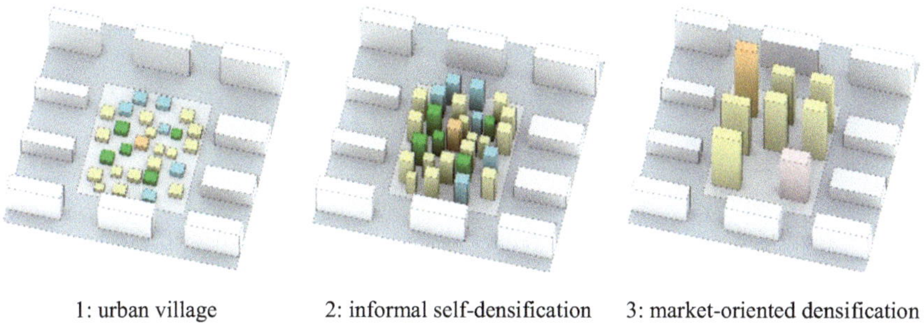

Fig. 5.1 High-density urban renewal process in Baishizhou. (*Source* Drawn by authors)

Phase 2: As urbanization accelerates, the region experiences informal densification, marked by self-built high-rises and increased population density. Environmental quality, living conditions, and infrastructure are often overlooked, resulting in a range of persistent issues.

Phase 3: With urban regeneration complete, the area enters a new phase of harmonious human-land relations. The approach focuses on demolition and new construction, achieving a balanced urban form. After regeneration, the region's population density and building structure reach an optimal level. This process maximizes land use potential, creating a sustainable urban morphology that is likely to remain for the long term.

In areas where land finance drives high land prices, this regeneration model remains prevalent. Notable examples include Baishizhou and Hubei in Shenzhen [18].

Baishizhou is one of Shenzhen's largest urban villages, home to the densest concentration of farmers' houses in the metropolitan area. Spanning only 0.6 square kilometers, it houses approximately 150,000 residents. The area is notorious for being disordered and unsanitary. with severe challenges in public safety, fire control, and health, complicating urban management. However, with ongoing urban development and transformation, regeneration in Baishizhou is now an urgent necessity (Fig. 5.2). The regeneration aims to improve the residential environment and quality of life for its residents. According to the urban regeneration plan, 480,000 square meters of land will be cleared to create a new commercial community with over 3.4 million square meters of construction area [3, 19].

5.2.1 Regeneration Update Content

According to the released Baishizhou draft plan [20], the land area of the regeneration unit of the Shahe five village urban regeneration project (i.e., Baishizhou old reconstruction) is about 480,100 m^2, the land area for demolition is about 459,500 m^2, and the development building area is about 303,800 m^2 (Fig. 5.3).

Fig. 5.2 Status before Baishizhou regeneration. (*Source* Author self-photography)

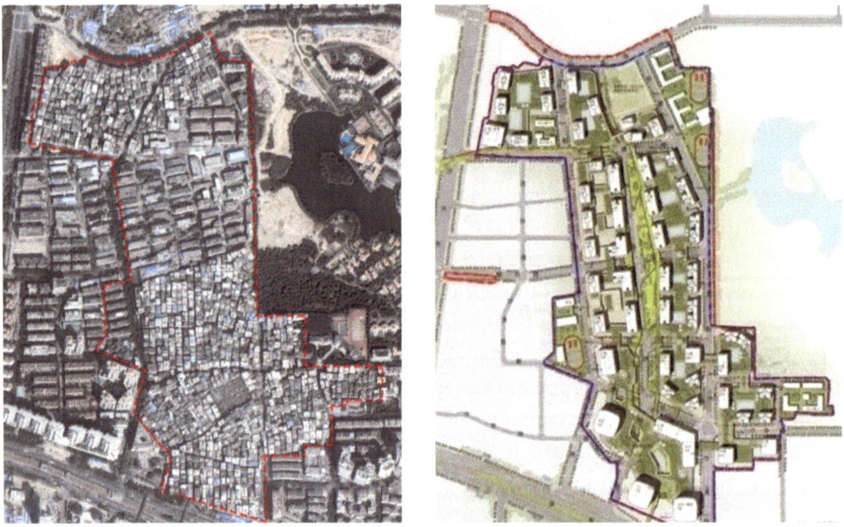

1.Baishizhou Satellite map 2008 2. Baishizhou Regulatory Masterplan 2020

Fig. 5.3 Comparison before and after urban regeneration. (*Source* From google map and Urban Planning, Land & Resource Commission of Shenzhen Municipality)

After reconstruction, the total floor area of the plot ratio is up to 3.4796 million m^2, including 1.25 million m^2 for residential buildings, 1.045 million m^2 for commercial, office and hotel industries, 1.12 million m^2 for business apartments, and 65,000 m^2 for public supporting facilities. The FAR is 11.5. The project consists of 31 residential buildings with 49–65 floors, 21 apartments, 3 super high-rise office buildings with 66–79 floors and 1 office building with 59 floors. In phase I, a 36-class nine-year

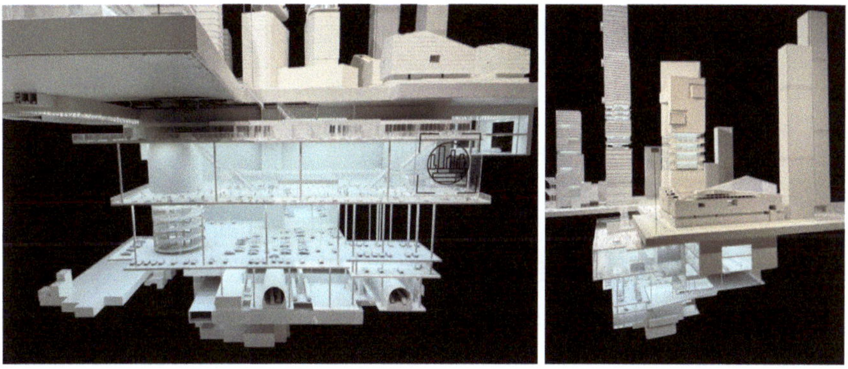

Fig. 5.4 Focus of high-density urban renewal—ground and underground. (*Source* Author self-photography and image processing)

school is planned. In phase II, a 27-class nine-year school, a 9-class kindergarten and a 12-class kindergarten are planned [21, 22].

5.2.2 Regeneration Update Subject

Baishizhou old reform is currently co-chaired by Baishizhou Industrial Co., Ltd., established by the Nanshan District government and village collective and Lvjing Real Estate from Hong Kong [23]. The district government mainly issued policies to maintain order. Baishizhou company is the main body of the old transformation. They have no real estate development qualification, so they cooperate with Lvjing. It is a mechanism of joint cooperation among the government, owners and enterprises.

5.2.3 Regeneration Benefits

Improved living conditions, increased various public facilities and supporting functions (Fig. 5.4), and reduced potential residential hazards. It has tapped the potential of the land economy and given full economic benefits to all parties [24].

5.2.4 Shortcomings of the Regeneration

Baishizhou has a very special meaning in Shenzhen for the continuation of place memory. It has retained the memory of many people [25, 26]. The daily life ecology it has created is also very fresh and valuable for Shenzhen, a city of immigrants.

The regeneration plan is reflected. In the preliminary plan for the old reconstruction of Shahe five villages in Baishizhou, one item is "village imprint andand cultural Baishizhou", which proposes to build a cultural park and preserve the important historical and cultural heritage of Baishizhou as the preservation of urban memory and culture, which is open to the public [27, 28].

5.3 Regeneration and Densification with Industrial and Infrastructure: Dafen Village

Densification is reflected in the economic benefits driven by land finance and the urban upgrading led by industry, which can further highlight the regional linkage advantages and the cluster advantages of diverse spaces. Dafen Village in Shenzhen is a typical case [29] (Fig. 5.5). A gathering area of famous agricultural and industrial paintings, and a village in the city [30].

Phase 1: At the initial stage, the area is mainly composed of old villages, supplemented by some self-built commercial and basic convenient functional buildings.

Phase 2: Accompany of the rapid urbanization period, the region has experienced partial urbanization by villagers themselves with vast informal densification. And Dafen village is located in the non-urban core area, and its densification degree is lower than downtown.

Phase 3: The vast densification is completed at this stage, and the regeneration process is transformed into the acupuncture model. According to the original oil painting industry agglomeration, the government provide industrial guidance and function improvement to promote industrial upgrading. New functional buildings are added to the original urban villages, and a prominent art museum has been built as a medium to publicize and encourage oil painting and relevant cultural and industrial development.

Dafen Village (with an area of 0.12 km^2) in Longgang District, Shenzhen. It was initially a remote village in the city. It has developed into an oil painting village with

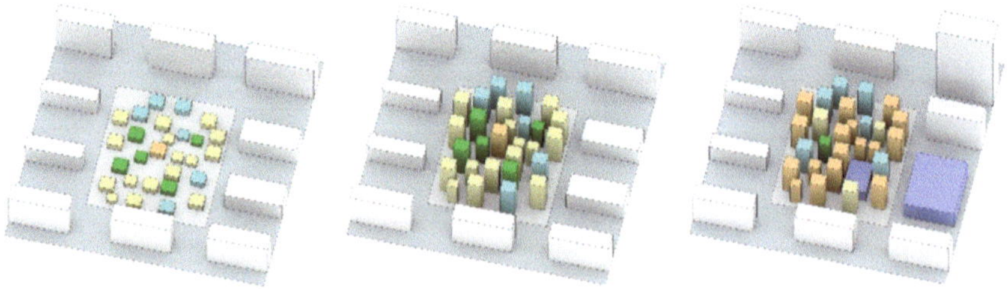

Fig. 5.5 High-density urban renewal process in Dafen Village. (*Source* Drawn by authors)

Fig. 5.6 Status before Baishizhou regeneration. (*Source* Author self-photography)

cultural value and brand effect. It was born through the gathering of cultural industries focusing on oil painting reproduction and has grown into an earlier gathering place of cultural and creative industries in Shenzhen [31]. With the industrial agglomeration and development, the urban regeneration of Dafen Village and its surrounding units, known as "the first village of Chinese oil painting", has become a critical project implemented in Longgang District at this stage (Fig. 5.6).

The updated content includes keeping the multi-storey, high-density compact small block shape spontaneously generated by the oil painting village in the urban regeneration. Space control is mainly implemented by optimizing urban buildings and public space [32]. In view of the current industrial model of art reproduction and original oil painting processing in the core area, the extension of the industrial chain is mainly through vertical upgrading and horizontal expansion [33].

The further development of the Dafen Village international art community requires the expansion of industrial space, mainly including the Longshan Industrial Zone on the south side of Busha Road and Mumianwan Village (about 0.67 km^2 in total) as the core area for the development of cultural and creative industries, and the resettlement of production, supporting facilities and population. At the same time, a new Dafen Village Art Exhibition Hall has been built as an important landmark in the leading area to undertake various exhibition and exchange activities [34].

Index before update: building density exceeds 40%. The total construction area is 139059 m^2, and the average number of floors is 4. Before the reconstruction, the total residential population of Dafen Village was about 10,200, and the per capita floor area was 9.21 m^2. The total floor area is 9.4 hectares, of which the main street is 7–8 m wide, the secondary street is 4–5 m wide, and the branch road is 1–2 m wide (Fig. 5.7).

After the update, the overall business format is commercial + residential. In the future, Dafen Oil Painting Village will be built into a "Dafen international art community" with four core functions: international oil painting trading center, enterprise headquarters base, industrial service platform and regional public center. In the

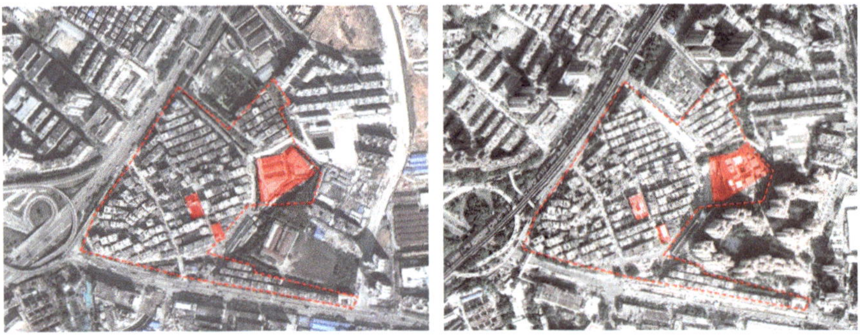

1. Dafen Village Satellite map 2002 2. Dafen Village Satellite map 2022

Fig. 5.7 Dafen Village Satellite map. (*Source* From google map)

future, no less than 52,248 m^2 of public interest land will be implemented within the reconstruction scope. Taking into account the new situation and new objectives of Dafen development, the development requirements of higher levels, as well as the direction of industrial regeneration, responding to the industrial upgrading and supporting space requirements of Dafen Oil Painting Industrial Park, improving the comprehensive service capacity of the area, and building Dafen into a comprehensive service center with four core functions: international oil painting trading center, enterprise headquarters base, industrial service platform and public center [35, 36].

Regeneration Development Oriented: guided by the government, industrial agglomeration development allowing the original urban villages to carry out industrial production in the form of land mixing, giving maximum industrial freedom [31].

5.3.1 Regeneration Benefits

It connects the overall industrial chain to form an industrial cluster, creating a complete supply chain system for Dafen Village and the surrounding intensive industrial space. Stimulate the vitality and creativity of individual industries in the original urban villages. Become the city's cultural card and attract many tourists [37].

5.4 Regeneration and Densification with Cultural Synergy: Nantou Ancient Town

Nantou ancient town was the former site of Bao'an County before the founding of Shenzhen, covering an area of about 385,000 square meters and a construction area of about 517,000 square meters. After historical events such as the relocation to the

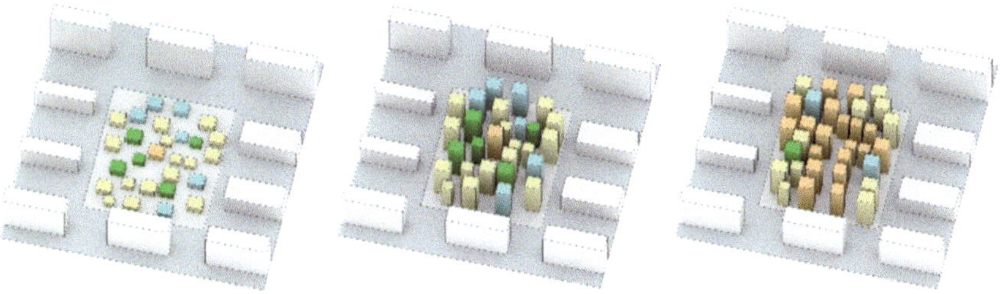

Fig. 5.8 High-density urban renewal process in Nantou ancient city. (*Source* Drawn by author)

sea in the late Qing Dynasty, the Japanese occupation during the Anti-Japanese War, and the rush to build villages in the city after the establishment of the special zone, the original style of the ancient town has been dramatically damaged [38].

With the continuous urbanization of Shenzhen [39, 40], the surrounding areas of Nantou ancient town are rapidly urbanized, and a complex pattern of "village in the city, city in the village" is formed around it [41]. The current resident population in the ancient town is about 30,000, of which about 90% are migrants. The high mobility of residents and the worry about the rising cost of living after the transformation make the residents and tenants have a low desire and ability to transform spontaneously. The traditional means of urban regeneration have failed. The activation and transformation of Nantou's ancient city need to get out of the dilemma of conventional thinking mode (Fig. 5.8).

Phase 1: As the birthplace of Shenzhen City, Nantou Ancient Town carries a large number of historical and cultural building types. Also, supplemented by some self-built commercial and basic convenient functional buildings.

Phase 2: Nantou ancient town, located in the city core area, has experienced partial urbanization by villagers with vast informal densification accompanied by a rapid urbanization period. Less concern about the quality of the environment, living, and infrastructure. This process brings many problems that are difficult to solve.

Phase 3: At this stage, the major street has completed the urban regeneration through the cooperation of the government, enterprises and owners. Due to this place's special historical and cultural background, the regeneration mainly carried out elevation renovation and functional renovation to preserve the traditional historical context. Through the scenarios creation and the injection of cultural symbols, a culturally represented commercial street with multi-experienced is formed as the sample project to drive the overall improvement of the whole region.

For areas where the site initially has a certain historical value, it needs a kind of cultural revival, cultural guidance, cultural benchmarking and appropriate densification intervention. Therefore, it is not only necessary to consider the architecture of its architectural design, but also need to consider the characteristics of the urban spatial network. The space efficiency it should play, the possibility of a series of functions and the sense of creating places through design. Mixing means a connection with

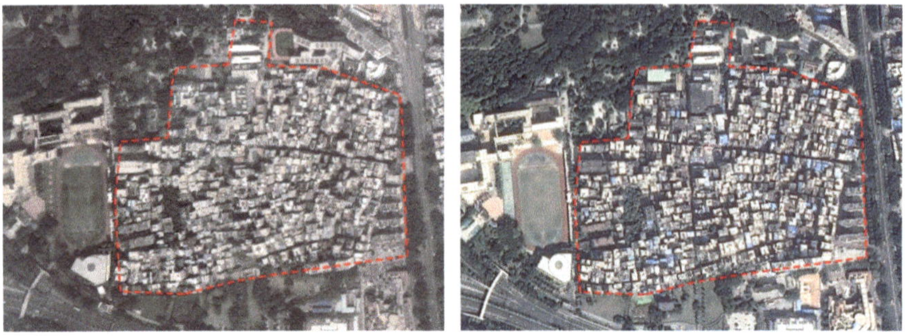

1. Nantou ancient city Satellite map 2010 2. Nantou ancient city Satellite map 2022

Fig. 5.9 High-density urban renewal process in Nantou ancient city. (*Source* From google Maps)

the future [42, 43]. Diversity and unity refer to the affiliation of functions under their formal language. Space itself is a kind of bearing, and the bearing content is the most fundamental. It is an important factor that causes people to pay attention to space. Therefore, the elasticity of content and the adaptability of content are very important.

Residents and tenants have a low desire and ability to transform spontaneously [44]. The traditional means of urban regeneration have failed, and the activation and transformation of Nantou's ancient city have stepped out of the dilemma of the conventional thinking model (Fig. 5.9).

It abandoned the controversial large-scale demolition and renovation and discussed "design + history and culture" to create a cultural and creative town.

Focusing on the reconstruction of buildings along the street and the replacement of business forms, it has been gradually updated in the form of "streets and Alleys" (Fig. 5.10). The revitalization of Nantou ancient city has been stimulated through the new creative vitality and humanistic quality. After the reconstruction, Nantou ancient town will be divided into four functional areas: historical nostalgia area, art and culture experience area, quality living area and cultural creativity area. Looking at the Nantou ancient city after the initial transformation [45], the demonstration section of the North South Street, which is more than 330 m long, has introduced diversified business forms such as catering, characteristic retail, cultural and creative space, public exhibition, etc.

We will protect and repair cultural relics and historical buildings, build characteristic historical and cultural blocks according to the Lingnan architectural style of the Ming and Qing Dynasties. Figure 5.11 introduce characteristic industries to drive the development of the area, and maximize the upgrading of the ancient town's infrastructure, community landscape environment and living quality.

Fig. 5.10 Nantou ancient city Street View. (*Source* Author self-photography)

Fig. 5.11 Nantou ancient town historical buildings. (*Source* Author self-photography)

5.4.1 Regeneration Implementation Body

The tripartite model of government leadership, enterprise implementation and villagers' participation. Nantou ancient city activation and regeneration project was officially launched, led by the government and implemented by Vanke Group (a private enterprise with government shares).

In march 2019, the government set up a working group on the protection and utilization of Nantou ancient city to rent all the buildings along the street for 15 to 20 years, introduce Vanke and other market players with relevant transformation and operation experience, and implement a unified leaseback building transformation and operation strategy.

5.4.2 Regeneration Benefits

Create a livable street with the most historical and cultural heritage in Shenzhen, and create a new business combination of "business + office + apartment" [46] (Fig. 5.12).

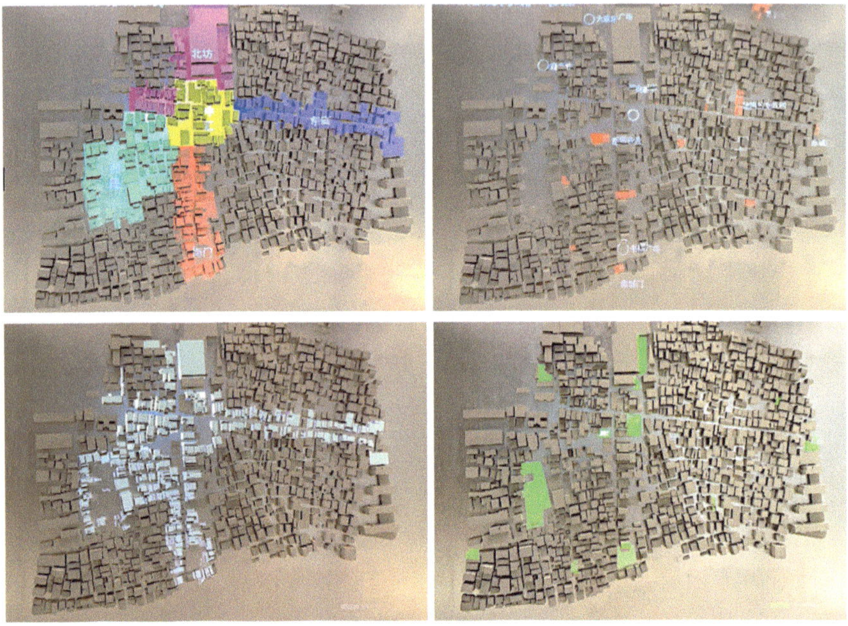

Fig. 5.12 Schematic diagram of space transformation of current cross street. (*Source* Author self-photography)

Create a paradigm benchmark and activate the continuous regeneration of Nantou's ancient city in the future. In the early stage, after the infrastructure construction is completed, the transformation around the main cross street will stimulate more market players to participate by activating some areas and transforming the node space.

The reconstruction of the social structure in the ancient city, the introduction of the design industry, the upgrading of commercial formats, the introduction of creative groups, and the formation of a mutually beneficial relationship with residents have become important driving forces for the revitalization of the ancient city [47].

After enriching the population structure, industrial structure and business structure, Nantou ancient city will gradually find its own regeneration mode in the process of change and diversification [48]. Nantou's ancient city will change from a low-end and residential village to a creative gathering place with complex cultural background and an organic ecosystem with cultural, commercial and residential symbiosis to achieve a rich and win–win situation in space, economy and culture. At the same time, while establishing the brand of Shenzhen as a cultural city, we will promote the coordinated development of surrounding areas [49].

5.5 Regeneration and Densification with Spatial Scenarios Creation: Shuiwei Ningmeng Apartment

The Shuiwei Ningmeng Apartment project is the first pilot project in Shenzhen to transform the "handshake building" in the urban village into a talent security housing community [50]. On the one hand, high housing prices and living costs have led to the loss of a large number of industrial talent, and there is a great demand for affordable housing for talent. On the other hand, due to the limited land resources and the difficulty of demolition and reconstruction of old areas, it is difficult for central urban areas to provide sufficient low-cost affordable housing for the migrant population in a short time [51]. The project uses a flexible market mechanism to solve the demand for affordable housing (Fig. 5.13).

Phase 1: At the initial stage, the area is mainly composed of old villages, supplemented by some self-built commercial and basic convenient functional buildings.

Phase 2: Accompany of the rapid urbanization period, the region has experienced partial urbanization by villagers with vast informal densification, ultra-high buildings, and population density. Less concern about the quality of the environment, living, and infrastructure. This process brings many problems that are difficult to solve.

Phase 3: At this stage, the region has completed the urban regeneration through cooperation with the government, enterprises, and house owners. Due to the special regeneration mode, it has become a sample project for the regeneration of urban villages. The high-rise residential areas are talented living apartments, and the commercial functions are located on the podium. Sky corridors are placed between the high-rise buildings as the medium of spatial connection, forming a desirable mixed-functional living place.

As the previous studies, the density of facilities, diversity of people, accessibility of public space, POI, commercial facilities, and a series of other factors will provide a vibrance in urban areas and promote abundant life scenarios. Considering relatively low-vibrance urban areas, with the limitation of capital investment and industrial

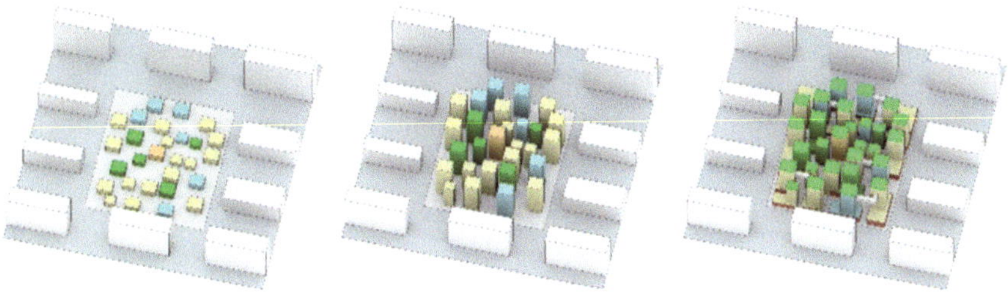

Fig. 5.13 High-density urban renewal process in Shuiwei Ningmeng Apartment project. (*Source* Drawn by author)

Ningmeng apartment Satellite map 2010 Ningmeng apartment Satellite map 2022

Fig. 5.14 Ningmeng apartment Satellite map. (*Source* Source from google map)

and culture upgrading, the densification strategy could adopt the creation of spatial scenarios.

The reconstruction area is located in Shuiwei village in the central region of Shenzhen, with a planned area of about 8000 square meters (Fig. 5.14). There are 35 unified farmers' buildings, 29 of which are transformed into 504 talent apartments, with a total construction area of 15,472 square meters [51, 52].

The improvement of the overall environmental quality has released substantial economic dividends. On the one hand, after the transformation, the 1–3 floors of shops operated and managed independently absorbed different types of industries such as culture, catering and entertainment, and the shop property was rented at an average price twice as high as the original price, rapidly increasing the property value. On the other hand, the rent and price of the independently renovated buildings have doubled, significantly increasing the commercial value of the whole village [53] (Fig. 5.15).

From self-built houses in a single urban village to a new life experience of co-construction and sharing, the transformation design maintains the original urban texture, architectural structure and the spatial scale of the characteristics of urban villages [54]. And by promoting firefighting, municipal supporting facilities and elevators, it has become a livable space in line with modern standards.

5.5.1 Regeneration Update Subject

The government-leased buildings, as the supply of talent housing, break through the traditional "government + enterprise + villagers" closed interest pattern [55], and

Fig. 5.15 Ningmeng apartment regeneration achievements. (*Source* Author self-photography)

comprehensively plan the living rights and interests of the migrant population. The apartment project was designed, constructed, operated and managed by Shenye land Co., Ltd. (a private enterprise).

The transfer of the right of use from the villagers to the village collective. In 2014, Shuiwei collective Co., Ltd. contacted more than 70 owners of 35 private residences that had moved overseas, hoping to obtain the right to use the property through a long-term lease.

Transfer of the right of use to the developer. Early in 2017, the Futian District government announced that Shenye Group Co., Ltd. would join the Ningmeng apartment renovation project, and the reconstructed housing would be included in the public housing of the Futian District.

Project approval and construction. In March 2017, the project entered the construction approval stage. Shenye invested 40million yuan in upgrading the internal layout and external infrastructure of the 29 farmers' houses to improve their living safety, convenience, comfort and aesthetics [56].

Apartment allocation. In December 2017, after the project transformation, Shenye transferred the upgraded 504 Apartments (a total of 15,472 m^2) to Futian District Housing and Urban–Rural Development Bureau for inclusion in the affordable government housing system.

5.5.2 *Regeneration Benefits*

"Do not push down and start again", preserve the texture, context and collective memory of the city, and inject new vitality and value into the old village.

It is discussed through the combination of old buildings and new culture to create a platform, trigger the participation and blending of residents in the old and new communities, and activate the old community.

It has solved and met the housing needs of young talents, improved the environment of urban villages, and quickly opened more possibilities for the transformation of old houses in urban villages.

The upgrading mode of the Ningmeng apartment is obviously different from the traditional urban village demolition reconstruction mode regarding social and economic achievements [57]. First, the demolition reconstruction method establishes clear formal property rights for buildings and land, while the Ningmeng apartment project retains the original informal property rights. Secondly, in the process of demolition, the original property right of the property is usually redistributed among the villagers, developers and the public. In contrast, in the process of upgrading, only the use right is transferred, and the property owner is still owned by the original oblige. Third, the building structure in the transformation and upgrading mode remains unchanged, while the demolition reconstruction mode is completely demolished and the new building structure is rebuilt. Fourth, the demolition and reconstruction mode gives the original owners a one-time compensation. In sum, the upgrading and reconstruction mode protects the long-term interests of the owners through stable and increasing rental income.

References

1. Meng L, Sun Y, Zhao S (2020) Comparing the spatial and temporal dynamics of urban expansion in Guangzhou and Shenzhen from 1975 to 2015: a case study of pioneer cities in china's rapid urbanization. Land Use Policy 97:104753
2. Zhang X, Chen J, Wang H, Yang D (2025) From policy synergy to equitable greenspace: unveiling the multifaceted effects of regional cooperation upon urban greenspace exposure inequality in China's megacity-regions. Appl Geogr 174:103472. https://doi.org/10.1016/j.apgeog.2024.103472
3. Chen J, Zhao X, Wang H, Yan J, Yang D, Xie K (2024) Portraying heritage corridor dynamics and cultivating conservation strategies based on environment spatial model: an integration of multi-source data and image semantic segmentation. Herit Sci 12:419. https://doi.org/10.1186/s40494-024-01497-7
4. Chen J, Li P, Wang H (2025) Assessment of influential factors on commute and life satisfaction in a historic campus-adjacent environment: evidence from a comparison study of Twin Cities. J Urban Plan Dev 151:4024065. https://doi.org/10.1061/JUPDDM.UPENG-5171
5. Shenzhen daily SZ ranks 9th in global urban competitiveness
6. Yang G, Zhou C, Jin W (2020) Integration of migrant workers: differentiation among three rural migrant enclaves in Shenzhen. Cities 96:102453
7. Tan X, Altrock U, Schoon S, Zhao J (2019) Localized place-making and the knowledge-based regeneration strategies-the case of Xiasha village in Shenzhen. Habitat Int 83:73–84
8. Pan W, Du J (2021) Towards sustainable urban transition: a critical review of strategies and policies of urban village renewal in Shenzhen, China. Land Use Policy 111:105744. https://doi.org/10.1016/j.landusepol.2021.105744
9. Chen J, Ren K, Li P, Wang H, Zhou P (2024) Toward effective urban regeneration post-covid-19: urban vitality assessment to evaluate people preferences and place settings integrating LBSNs and POI. Environ Dev Sustain. https://doi.org/10.1007/s10668-024-05378-8
10. Zhao M, Liu N, Chen J, Wang D, Li P, Yang D, Zhou P (2024) Navigating post-COVID-19 social–spatial inequity: unravelling the nexus between community conditions, social perception, and spatial differentiation. Land 13
11. Song H, Chen J, Li P (2025) Decoding the cultural heritage tourism landscape and visitor crowding behavior from the multidimensional embodied perspective: insights from Chinese classical gardens. Tour Manag 110:105180. https://doi.org/10.1016/j.tourman.2025.105180

12. Hu J, Chen J, Li P, Yan J, Wang H (2024) Systematic review of socially sustainable and community regeneration: research traits, focal points, and future trajectories. Build 14:881. https://doi.org/10.3390/buildings14040881
13. Blackwell A (2021) The Shenzhen experiment: the story of China's instant city, by Juan Du
14. Tian L, Yan Y, Wu Y, Shao L (2020) Breaking the land monopoly: can collective land reform alleviate the housing shortage in China's mega-cities? Cities 106:102878
15. Herrmann Pillath C (2018) Is Shenzhen emerging as a global city with distinct cultural roots? Rep Pilot Study Prep China Dev Inst, Shenzhen 57
16. Chcn J, Gan W, Liu N, Li P, Wang H, Zhao X, Yang D (2024) Community quality evaluation for socially sustainable regeneration: a study using multi-sourced geospatial data and AI-based image semantic segmentation. ISPRS Int J Geo-Inf 13(5):167. https://doi.org/10.3390/ijgi13050167
17. Zhao X, Chen J, Li J, Wang H, Zhang X, Yu F (2025) Unraveling the renewal priority of urban heritage communities via macro-micro dimensional assessment-a case study of Nanjing city, China. Sustain Cities Soc 124:106317
18. O'Donnell MA, Bach J (2021) Reclaiming the new, remaking the local: Shenzhen at 40. China Perspect 2021:71–75
19. Chen J, Li P, Wang H (2024) MSPA-based collaborative evaluation and planning governance of territory and natural environment in metropolitan agglomerations—a case study of EGIDDZ in Yangtze river delta. J Resour Ecol 15:1263–1273. https://doi.org/10.5814/j.issn.1674-764x.2024.05.014
20. Backholm J (2019) Urban redevelopment in Shenzhen, China neoliberal urbanism, gentrification, and everyday life in Baishizhou urban village
21. Chen J, Li P, Lei Y, Zhang Y, Lai C, Chen B, Liu J, Schnabel MA (2025) Leveraging augmented reality for historic streetscape regeneration decisionmaking: a big and small data approach with social media and stakeholder participation integration. Cities. https://doi.org/10.1016/j.cities.2025.106214
22. Yang N, Chen J, Ban L, Li P, Wang H (2024) Pre-planning and post-evaluation approaches to sustainable vernacular architectural practice: a research-by-design study to building renovation in Shangri-La's Shanpian House, China. Sustain 16:9568
23. Verbeelen BC (2010) The collaboration of actors in urban village redevelopment projects in Shenzhen, China. Int New T Inst 1–77
24. Shenzhen GeoKey group Shenzhen "old aircraft carrier" ushers in new progress in construction: phase I of Baishizhou urban renewal project has come out of the ground
25. Chen J, Wang H, Yang Z, Li P, Ma G, Zhao X (2023) Comparative spatial vitality evaluation of traditional settlements based on SUF: taking Anren ancient town's urban design as an example. Sustain 15:8178
26. Chen J, Pellegrini P, Yang Z, Wang H (2023) Strategies for sustainable urban renewal: community-scale GIS-based analysis for densification decision making. Sustain 15:7901
27. Chenjing Fan, Yueying Gu, Chenxi Jin, Zhihao Yao, Runhan Liu (2025) Does urban renewal enhance urban vitality? A case study of Nanjing, China, using multidimensional data. Environ Plan B Urban Anal City Sci 23998083251347680. https://doi.org/10.1177/23998083251347682.
28. Ma G, Paola P, Wu H, Han H, Wang D, Chen J (2023) Impact of land-use mixing on the vitality of urban parks: evidence from big data analysis in Suzhou, Yangtze River Delta region, China. J Urban Plan Dev 149:4023045. https://doi.org/10.1061/JUPDDM.UPENG-4334
29. Li S, Cheng H, Wang J (2014) Making a cultural cluster in China: a study of Dafen Oil Painting Village, Shenzhen. Habitat Int 41:156–164
30. Wong WWY (2010) After the copy: Creativity, originality and the labor of appropriation: Dafen Village, Shenzhen, China (1989–2010)
31. Wang J, Li SM (2017) State territorialization, neoliberal governmentality: the remaking of Dafen Oil Painting Village, Shenzhen, China. Urban Geogr 38:708–728
32. Shi Y, Pan Y, Xu Z (2010) Urban complex design based on compact and intensive planning theory—a case study of oil painting Square in Dafen Village, Shenzhen. PlanS 26:64–67

33. Chen J, Tian W, Xu K, Pellegrini P (2022) Testing small-scale vitality measurement based on 5D model assessment with multi-source data: a resettlement community case in Suzhou. ISPRS Int J Geo-Inf 11:626. https://doi.org/10.3390/ijgi11120626
34. Shenzhen City Planning and Land Resources Committee (2010) Dafen in Shenzhen, the regeneration of an urban village
35. Zhang X, Lu Y, Xu Y, Zhou C, Zou Y (2024) Governing regional inequality through regional cooperation? A case study of the Guangdong-Hong Kong-Macau greater bay area. Appl Geogr 162:103135. https://doi.org/10.1016/j.apgeog.2023.103135
36. Zhang X, Zhou Z, Song Y (2024) Contested geopolitics of distribution in the city-region-building: promoting enclave industrial parks in the Pearl River Delta, China. Trans Inst Br Geogr e12697
37. Chen S (2021) Hoping amidst precarity: Changing modes of hope in the vicissitudes of a Chinese art village. Sociol Rev 69:1277–1292
38. Meng Y (2018) Village/City: Coexistence and regeneration in Nantou, Shenzhen. Time + Archit 3:58–64
39. Nanshan District People's Government of Shenzhen. Implementation plan for Nantou ancient city regeneration project ("butterfly plan") [R] (Document No.: NFFG [2020] No. 15). Shenzhen: Official Gazette, 2020-09-02. http://www.szns.gov.cn. Accessed 30 June 2025
40. Li S, Qu F (2022) Preserving authenticity in urban regeneration: a framework for the new definition from the perspective of multi-subject stakeholders—a case study of Nantou in Shenzhen, China. Int J Environ Res Public Health 19:9135
41. Zhang X, Sun Y, Luan X (2023) Regulation flexibility and legitimacy building in governing intercity railways: the polymorphous role of the Chinese provincial government. J Urban Aff 1–16
42. Zhang X, Shen J, Sun Y, Zhou C, Yang Y (2023) Formation of city regions from bottom-up initiatives: investigating coalitional developmentalism in the Pearl River Delta. Ann Am Assoc Geogr 113:700–716
43. Zhang X, Chen Y, Xu Y, Yue W, Tang C (2023) Does regional cooperation constrain urban sprawl? evidence from the Guangdong-Hong Kong-Macao greater bay area. Landsc Urban Plan 235:104742
44. Minsheng weekly (2021) Shenzhen's Nantou ancient city: a model of cultural regeneration [N]. (6):28–31. ISSN 2095-4501. http://paper.people.com.cn. Accessed 30 June 2025
45. Meng Y, Lin Y, Rao E (2017) Curator Nantou: A sample of city/village symbiosis and rebirth. Time + Archit 7
46. Ling J (2018) State territorialization, social territoriality and cultural zone: the remaking of Nantou Village in Shenzhen, China. City Univ, Hong Kong
47. Bian W (2021) Research on architecture color of transformation and renewal strategy in old district based on the background of urban rehabilitation: a case study of Shenzhen Nantou historic town. Mod Urban Res 000:73–79
48. Zhang X, Li Y, Zhou C, Luan X, Yuan F (2023) Rescaling of the land regime in the making of city-regions: a case study of China's Pearl River Delta. Urban Stud 60:483–500
49. Liu Y, Lv B (2018) Place making in urban rehabilitation of Nantou Ancient Town, Shenzhen. PlanS 34:7
50. Li B, Tong D, Wu Y, Li G (2021) Government-backed 'laundering of the Grey' in upgrading urban village properties: Ningmeng apartment project in Shuiwei village, Shenzhen, China. Prog Plann 146:100436
51. Pan W (2019) China Southern: Digital environments as geopolitical contact zones. Concordia Univ
52. Cai W (2019) Ningmeng talent apartment, Shenzhen, China. World Arch 134–137
53. Che Y (2019) Rebuilding of space in villages belong to the city under the background of shared-economy era—taking Shuiwei Ningmeng talent apartment in Shenzhen as an example. City House 6
54. Ni H, Lin J, Lin Y (2020) A research on the symbiotic design strategy combining youth apartments and hackerspaces. J Guangdong Univ Technol 37:98–104

55. China Young Planners Association (2018) Public-private-village collaboration in urban village renewal: case study of Ningmeng talent apartment, Shuiwei village, Shenzhen [R]. Shenzhen: Urban Planning Society of China. 45–59. https://doi.org/10.12345/cypa.2018.055
56. Pellegrini, Paola; Chen, Jinliu The Regeneration Path Not Taken Further: An Experiment in Urban Densification and State Entrepreneurialism in a Resettlement Neighborhood in Suzhou, China. Urban Stud. 2025, 00420980251365188, https://doi.org/10.1177/00420980251365188
57. Futian District Housing and Construction Bureau of Shenzhen. Project application report for Shuiwei Ningmeng talent apartment construction [R] (Approval No.: FTJSJ [2020] 037). Shenzhen: Futian government online, 2020-03-15. http://www.szft.gov.cn. Accessed 30 June 2025

Chapter 6
Densification Assessment Through Residents' Satisfaction in Nanhuan Community, Suzhou

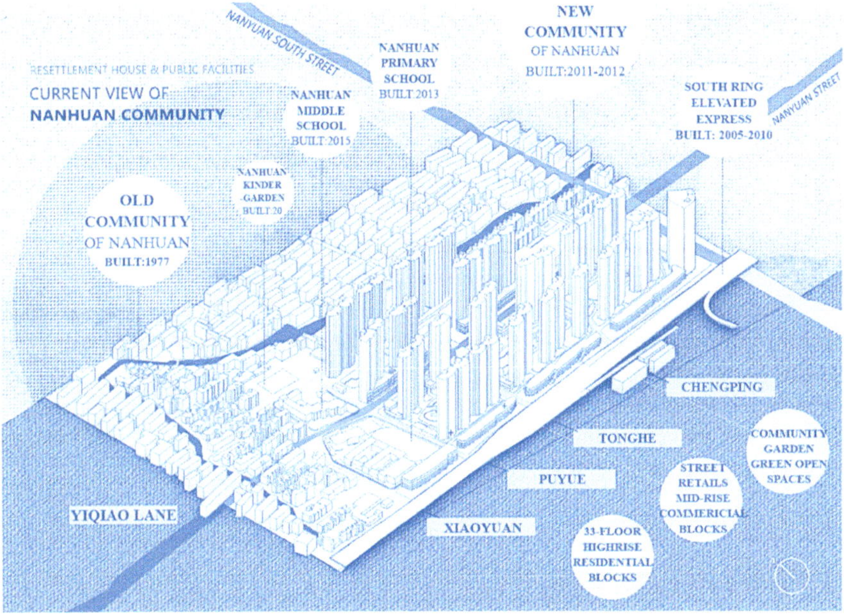

6.1 The New Role of Regeneration in Chinese Urbanization

The 14th five-year plan of China (2020–2025) assigns a relevant role to urban regeneration in the process of sustainable and high-quality urbanization [1, 2]. Urban regeneration is officially promoted for the first time by the central government to use land resources efficiently, create a better living environment, and improve the city image [3, 4]. The intentions of the 14th five-year plan urban regeneration include a range of actions: from demolition with reconstruction to partial renovation, to

J. Chen and P. Pellegrini, *Densification as a Design Tool for Sustainable Urban Regeneration*, Urban Sustainability, https://doi.org/10.1007/978-981-95-1653-7_6

comprehensive renovation [5]. These very recent guidelines align China with the directions for sustainable development, that is the main goal of the UN-Habitat [6–9].

Beyond the basic improvements of the building stock, the idea of regeneration in China brings about two emerging topics: the increase of population density in urban areas and a new approach to social sustainability [10, 11]. In fact, intending to increase the urbanization rate, which is a key national priority, and under the pressure of constant urban immigration, the most competitive cities in China are increasing the existing FAR: on one side to make the investment in urban regeneration profitable, on the other side to use land more efficiently [4, 12].

Different from the past, the social sustainability of each transformation is increasingly considered, because not only growth must be achieved, but sustainable growth according to the New Type of Urbanization launched in 2014 with full consideration of residents' needs and satisfaction [13–15].

This chapter addresses the topic of residents' satisfaction with the regeneration and densification of their resettlement neighborhoods built before 2000 in the city of Suzhou, Yangtze River Delta region, China [16, 17]. The chapter focuses on residents' satisfaction and aims to answer three main research questions: How has the regeneration impacted the residents' satisfaction? What elements and changes in the open space and buildings have significantly impacted residents' satisfaction? What role has the densification played in the formation of the residents' perception of the regenerated community [12, 18]?

These research questions connect space and satisfaction and aim to discover if the regeneration project, which has transformed the existing neighborhood into a high-density one, has increased residents' satisfaction or not [19]. The chapter assumes residents' satisfaction as one element to measure the success of the regeneration and the social sustainability of the transformation process. The chapter wants to contribute to the discussion about the ongoing Chinese transition process into high-quality residential communities and the general negative assessment of high-density levels in the built environment [20].

6.2 Literature Review

6.2.1 Old Resettlement Communities: A Critical Part of Chinese Cities

The resettlement communities (安置社区) are neighborhoods where people who had the–r house demolished—farmers or inhabitants of old urban centers o–dilapidated areas—are relocated according to specific policies [21, 22], which are meant to provide sufficient construction land for urbanization, improve land-use efficiency, offer the residents both an improvement of their houses and compensation [3, 16]. The resettlement was usually financed by local governments and given a low budget:

planning was as rapid as possible, the design largely standardized, the construction planned to last not more than 30 years; given these premises decaying installations, outdated space utilization, ineffective open space distribution, inadequate supporting public facilities, and infrastructure breakdowns are frequent [5].

The obsolescence of the structures, the poor maintenance, and the missing facilities, combined with the enrichment of Chinese society, make the resettlement communities built before 2000 a critical part of Chinese cities [23]. Under these conditions, the regeneration of resettlement communities should start from an overall re-thinking of the living environment, including residential units, FAR, dimensions, distribution, allowed uses of the open spaces, and supporting facilities [24, 25].

In China, the old resettlement communities are one of the main targets of regeneration for several reasons related to the quantity and the quality of the stock: (1) The amount to a significant proportion of the existing housing stock; in fact, nearly 100 million households experienced housing demolition and relocation from 1995 to 2014 [26]. (2) Several resettlement communities offer a low-quality built environment and few public services, and therefore need the upgrading of both the indoor features and the outdoor public spaces, basic living infrastructures, and supporting facilities [27, 28].

According to the 14th five-year plan and the related guidelines, the regeneration must target especially the communities built before 2000 and produce a living environment better suited to the newly achieved wealth of a large part of the Chinese population, that is to say, the middle class which the central government wants to increase. The improvement is consistent with the goal of the moderately prosperous society" (小康社会) promoted for the whole country by the central government [29].

6.2.2 Quality of Living Environment, Social Sustainability and Residents' Satisfaction

Social sustainability is one of the pillars of sustainable development, because it satisfies basic human needs, improves life, and ensures social justice [30]. The quality of the living environment is essential in determining the quality of life because it impacts the fulfilment of needs, everyday life, and social justice. Improving the quality of life (QoL) is a key component in enhancing social sustainability at the community level [31, 32]. QoL implies subjective life satisfaction [33]. The measurement of QoL is based on residents' satisfaction, with the investigation of subjective feelings, then promoting comprehensive consideration of living well-being and social balance, and ensuring social justice [34].

At the neighborhood level, a socially sustainable community can be defined as a place where people want to live and work now and in the future, because: (1) it is a long-term viable setting for human interaction, communication, and culture development. (2) it meets the diverse needs of current and future residents, is environmentally sensitive, and contributes to a high QoL. (3) it is safe, inclusive, well-planned, well-constructed, and managed, and provides equal opportunities and good services to all.

According to numerous scholars, residential evaluation, or "satisfaction" with the neighborhood where one lives, is an important indicator to measure the quality of the living environment. Analyzing residents' evaluation has been identified as a primary method to check the achievements of planners and basic government rules [35].

Since the 13th national five-year plan (2015–2019), social sustainability is officially recognized in China along with ecological sustainability as a significant factor in achieving high-quality urbanization and people-oriented development [36]. The achievement of a "harmonious society" (和谐社会) in the fast-developing Chinese cities is connected to residents' satisfaction, and the factors affecting residents' satisfaction are receiving unprecedented attention in China [37–40]. Moreover, even though there is no specific regulation about people's participation in planning and design activities, the Chinese government regeneration policy states the importance of residents' participation in regeneration projects to understand their needs and priorities and deal with the conflicts which often raise between residents and local governments for the transformation decision-making [41, 42].

In China, specific studies to measure residents' satisfaction started relatively late if the global context is considered and a variety of approaches coexists in current research [43, 44]. Some studies used quantitative and statistical analysis methods to define the influencing factors; others used qualitative comparison and descriptive analysis [45]. According to existing studies, the residents' satisfaction with large-scale resettlement communities can be analyzed using a comprehensive and multi-level indicator system that considers objective and subjective factors: (1) objective factors include the characteristics of the residential unit, the surrounding environment, and infrastructure [46]. (2) subjective factors include personal and family characteristics, income level, house ownership, relocation, and compensation for relocation [35, 47].

When selecting the evaluation indicators, we selected a broad range of indicators to represent all the aspects that are closely related to people's behavior and life, such as property, pedestrian, greening, cultural symbols, etc., and achieve as complete a description as possible. The factors can be grouped into 3 main categories: (1) housing conditions, such as building quality, building area, building age, building orientation, lighting, and ventilation [48]. (2) individual attributes, such as age, education, family structure, economic level, house ownership [49]. (3) context, such as community management, supporting infrastructure, transportation convenience, surrounding environmental conditions, neighbor's relationships [49, 50].

6.3 Research Method

6.3.1 The Case Study

Several studies in China have focused on the impact of various spatial characteristics of some communities on residents' satisfaction [51–53], but very few studies analyzed the residents' satisfaction before and after the transformation of their living environment [54, 55]. This comparative study is possible in the case of Nanhuan Village (南环新村), in Suzhou, Yangtze River Delta area, where one part of the community was transformed in morphology, building types, and density, and one part remained as originally designed in the '70s (Figs. 6.1 and 6.2). Nanhuan Village, in fact, was built between the late '70s and the early '80s to relocate over 6000 inhabitants from Suzhou old town, which was being demolished (first relocation in the first stage of city modernization). The resettlement community was designed "as usual" in this kind of operation: a compact sequence of about 150 almost identical 4–6 floors residential buildings-oriented east–west with no open space beyond the ones strictly needed between adjacent buildings. In 2010 approximately 80% of the community was demolished for its dangerous obsolescence and low living standards—such as foundation subsidence, house tilt, balcony collapse, road damage, and inadequate supporting facilities—and replaced by a completely different urban morphology and building types (second relocation for regeneration in a mature phase of urbanization); the construction of 21 high-rises of 22 to 32 floors increased the FAR from 1.5 to 2.6.

In May 2010, the demolition of 118 buildings on a total built area of 325,000 square meters began; in June 2011, the agreement for resettlement and compensation of 4778 households was settled; in June 2013, the New Village was completed and 97% of the original residents moved back. The New Village required an investment of more than 2.3 billion yuan to build 559,800 square meters for 5137 households, 2 schools, and one community center, commercial spaces in one mall, and multi-functional commercial streets. Now, approximately 15,000 residents live in the Nanhuan whole community.

Fig. 6.1 Aerial view of the two parts of Nanhuan Village. Photos by the authors, 2023. (On the left, the buildings of the original resettlement village, on the right, the area transformed by high-rises)

Fig. 6.2 Nanhuan New Village master plan, first version. (The old part of Nanhuan is west of the new part; the new schools are between the old and the new part)

The Nanhuan case preceded the current national guidelines for regeneration by 10 years and attracted attention for its innovative proposal: (1) it is the first regeneration of an existing resettlement neighborhood planned, financed, and realized by the local government in Suzhou; the project was included in the government's annual list of crucial tasks and it was meant to be exemplary for communities in similar conditions. (2) it included densification and residents' relocation in situ. (3) It combined high-rises with small compact blocks and mixed uses, an unusual solution where mono-functional superblocks are the most frequent condition.

As the transformation involved only one part of the neighborhood, Nanhuan Village allows a direct comparison of the residents' satisfaction with the original solution and with the high-density one, the New Village [56, 57].

6.3.2 *Research Steps*

The research on Nanhuan residents' satisfaction adopted a comprehensive and multi-level indicator system, which includes objective and subjective factors to (1) identify and analyze the factors and indicators, which affect residents' satisfaction. (2) compare the residents' satisfaction with the old part of Nanhuan Village and with the new part to understand the changes due to the regeneration.

The steps of the investigation were: (1) field research, data collection, analysis of residents' satisfaction: interviews with experts and stakeholders involved in the realization of the old and New Village; field observation and survey with questionnaires distributed to the residents of both parts; evaluation of residents' satisfaction levels. (2) analysis of the influencing factors and results: identification of the factors and indicators which influence residents' satisfaction, applying an exploratory factor analysis (EFA) and a SEM. In considering both objective and subjective factors, the analysis has not applied a single-equation model because the influence of objective factors may be underestimated [58].

6.4 Data Analysis and Results

6.4.1 Field Research and Questionnaire

A preliminary field investigation was carried out and people with different expertise and involvement in the community were interviewed to gain an in-depth understanding of the current situation, its background, and what influencing factors of satisfaction are identified by them [53]; they are the management of the community, the head of the planning and design institute who designed the Master Plan of the New Village, 3 members of the neighborhood committee, 10 residents.

The questionnaire adapts the indicators most frequently listed in the relevant literature to the information collected in the preliminary investigation: (1) Personal identity, Lee et al. concluded that individual identity, including gender, education, income, and race influences residents' satisfaction. Parkes et al. believed that low-income, poor jobs, and tenants have a negative impact on residents' satisfaction [59]. (2) Living space, Mccrea et al. sorted out that the size of the house, the degree of regional pollution, the age, and the status of the traffic peers are all the factors affecting the overall residents' satisfaction. Mohit and Azim believed that the size and housing conditions, and the pedestrian passage were the main variables of the residents' satisfaction [60]. (3) Services and facilities, Yi et al. regarded facilities, transportation, and property as indicators of residents' satisfaction. Phillips and others believed that the elderly facilities and security are relatively significant. Wang and others believed that parking, service facilities, and entrance and exit traffic have a great impact on Residents' satisfaction [49, 50]. (4) Social network, Mohit et al. believed that community residents' relations and rest space have a great impact on residents' satisfaction, and shopping preferences also have an important impact [61].

Then we integrated the existing literature to form the index system with 23 indicators in Table 6.1.

The questionnaire was mainly sampled according to the proportion of the community population (1/50), and questionnaire sample collection as conducted by means of convenience sampling. Convenience sampling means a non-probabilistic sampling method of randomly selecting respondents at a specific time and within a specific community area. A total of 308 questionnaires were conducted face to face in the public spaces of both the old and new parts of the Village on weekdays between

Table 6.1 List of the indicators in the questionnaire

Satisfaction indicators	Secondary indicators
Personal information (A)	Age (A1)
	Gender (A2)
	Household registration (A3)
	Occupation (A4)
	Current residence (A5)
	Residence before relocation (A6)
	Income (A7)
Living space (B)	House type (B1)
	Dwelling area (B2)
	Person/each unit (B3)
	Building quality (B4)
	Air quality (B5)
	Pedestrian safety (B6)
Community convenience degree (C)	Parking convenience (C1)
	Supporting facility convenience (C2)
	Aging care service (C3)
	Education (C4)
	Public transport convenience (metro/bus) (C5)
	Entrance convenience (C6)
	Property management (C7)
Social network (D)	Green space landscape (D1)
	Neighbor interaction (D2)
	Shopping preference (D3)
Satisfaction	Residents' satisfaction degree

8–10 am and 5–7 pm in summer 2021; 295 questionnaires were completed with an effective rate of 95.8% (Fig. 6.2). Team members inquired the residents randomly in the Nanhuan New Village public areas and asked whether they were willing to do a questionnaire survey. After obtaining their consent, the process of questionnaire collection was started. Team members explained the goal of the questionnaire and the available choices to the participants to ensure they answered accurately. The team obtained the approval of the survey activity from the neighborhood managing committee, and with their support, also some residential units were visited [62, 63].

6.4.2 Analysis of the Level of Residents' Satisfaction

The research analyzed and classified in a range of values 1–5 the answers in the questionnaires to define the main characteristics of the residents in Nanhuan and to evaluate the residents' satisfaction (Tables 6.2, 6.3, and 6.4). The main findings are as follows.

Table 6.2 Results of the individual information in the questionnaire of the whole Nanhuan Village (selection)

Statistical variables	Category	Count	%
Age	20–45 years old	130	43.20%
	46–59 years old	61	20.30%
	Over 60	110	36.50%
Gender	Male	173	57.50%
	Female	128	42.50%
Household registration (Hukou 户口状态)	Suzhou	204	67.80%
	Out of town	97	32.20%
Occupation	Unemployed	147	48.80%
	Private enterprise	81	26.90%
	State-owned enterprise	73	24.20%
Current residence	Old district	78	25.90%
	New District	217	72.10%
Residence before relocation	Nanhuan	127	42.20%
	Suzhou	102	33.90%
	Other city	72	23.90%
Income	No income	4	1.30%
	Below 5000 yuan	38	12.60%
	Between 5000 and 7000 yuan	52	17.30%
	Between 7000 and 10,000 yuan	60	19.90%
	Between 10,000 and 15,000 yuan	76	25.20%
	Between 15,000 and 20,000 yuan	53	17.60%
	More than 20,000 yuan	18	6.00%

Table 6.3 Residents' satisfaction degree from the questionnaire output

	Category	Count	Percentage
Satisfaction	1 score	3	1%
	2 score	20	6.6%
	3 score	134	44.5%
	4 score	139	46.2%
	5 score	5	1.7%

Table 6.4 Mean values for satisfaction evaluation

Primary indicators	Secondary indicators	Mean value for secondary indicators (1–5)			Standard deviation		
		Total	Old	New	Total	Old	New
Living space	Number of rooms	2.38	2.21	2.44	0.719	0.543	0.756
	Dwelling area	2.32	2.24	2.33	0.968	0.983	0.952
	People/each unit	2.63	2.54	2.65	1.062	0.935	1.104
	Building quality	3.29	3.31	3.28	0.788	0.827	0.781
	Air quality	3.20	3.27	3.18	1.145	1.170	1.143
	Property management	3.11	3.04	3.12	0.831	0.813	0.841
	Pedestrian safety	2.96	2.97	2.96	0.981	0.911	1.001
Community convenience degree (CCD)	Parking convenience	1.68	1.38	1.79	1.526	1.589	1.488
	Supporting facility convenience	3.62	3.53	3.66	0.954	0.990	0.950
	Aging care service	1.59	1.71	1.54	0.964	0.982	0.962
	Education	1.66	1.64	1.67	0.473	0.483	0.470
	Public transport convenience	3.81	3.88	3.78	0.866	0.837	0.879
	Entrance convenience	3.46	3.38	3.49	1.053	1.154	0.996
	Improvement after regeneration	2.36	2.32	2.36	1.578	1.827	1.488
Social network	Green space landscape	3.36	2.32	3.73	0.807	0.634	0.476
	Neighbour interaction	1.57	1.51	1.61	1.157	1.170	1.162
	Shopping preference (online)	1.62	1.64	1.62	0.512	0.509	0.514

As the collection of information obtained by the questionnaires (Table 6.2) highlights, on average, the old and the new parts have similar inhabitants in age, income, occupation. 66.4% of the residents have an education level lower than high school, while in Suzhou it is 60.6% and in the Jiangsu province is 73% in 2020. This data means the neighborhood is mostly occupied by a low-skill labor force (Table 6.2).

Both parts of the village remained attractive over the years because the original residents were largely substituted (Table 6.2): while 98% of the residents who had their house demolished returned to the Nanhuan New Village in 2014, today only 40% lived in Nanhuan before relocation. The neighborhood attracts immigrants: more than 30% are not Suzhou citizens (the hukou status, 户口状态); the New Village accommodates 10% more immigrants than the old part. The presence of this floating population, very likely low-skilled workers, means that the rent prices in the neighborhood are affordable.

The neighborhood also accommodates middle-class residents: in fact, the most represented group, 25%, declared the highest income, 10,000 to 15,000 RMB a month, much higher than the average income declared in Suzhou, 5900 RMB, and in the Jiangsu province, 4400 RMB in 2020 (Table 6.2). As the select results of questionaries, the indicators do not show radical differences between the old and new part of the Nanhuan community; the most relevant are: the number of immigrants is 8.5% higher in the new part; the number of unemployed or retired are 5.6% higher in the old part; the apartments are larger in the new part; the old part is inhabited by families with only two components 9% times more than the new one; residents feel much safer in the old part; residents of the new part are more satisfied with the green landscape.

From the questionnaire analysis result, it can be said that most residents share the same opinion: 46.2% of the residents declares to be satisfied, score 3, 44.5% declares to be satisfied, score 4 (Table 6.3).

The questionnaire samples' residents' satisfaction result between new and old Nanhuan community. Beyond this overall appreciation, which can be influenced by the Chinese cultural milieu, the results show that the different design of the two parts has an impact on the level of residential satisfaction. The New Village scores slightly better. The indicators that are most relevant to assess the result of the transformation of the old neighborhood into the new one score well, the ones which urged the process of transformation, such as building quality, green space, supporting facility, public transport. The assessment of the regeneration is positive, as shown for example by the average values of the primary indicators "living space," "community convenience," and "social network" are higher than the medium score in both parts (Table 6.4). Within the scope of this research, which compares the condition before and after the regeneration of the neighborhood, it is particularly relevant that "residence change" and "living condition" have no significant impact on the overall satisfaction, because it means that the residents' assessment is not impacted by the radical change in their living environment due to the demolition and rebuilding of New Nanhuan.

The results show how the old residential environment is still valued (as an example because it is safer and better connected to public transport), and the regeneration has improved the facilities for both parts not only the new part.

Residents of both parts gave the lowest satisfaction scores to neighbor's interaction, shopping facilities, and parking. The complaint about parking is justified because the design underestimated the ratio of parking lot per household: only 0.25, while in 2020 there was 0.82 car in every household in Suzhou on average; in fact, the regulation today requires 0.8 parking lot per unit.

It must be noted that the "building quality", which includes the management of the compound, is more appreciated by the residents of thc old part simply because they have no elevators and are not impacted by their management. Similarly, the green landscape is assessed much better by the residents of the new part because the old part has almost no green areas.

Table 6.5 Exploratory factor analysis of influencing factors

Factor name	Included variables (factor loading)
Individual attributes	Age (0.853), occupation (−0.781), shopping preference (0.698)
Community environment	Property management (0.762), building quality (0.674), entrance convenience (0.557), air quality (0.517)
Greening level	Green space landscape (0.916), residence Hukou (0.909)
Living condition	Number of rooms (0.794), dwelling area (0.787)
Residence life	Education (−0.853), people/each unit (0.762)
Self-conditioned cognition	Income (0.852), pedestrian safety (−0.640), gender (−0.463)
Residence change	Improvement after regeneration (0.747), residence Hukou (0.616), residence before moving (0.567)
Supporting facility	Supporting facility convenience (0.732)
Neighbor interaction	Neighbor interaction (0.732)

6.5 Analysis of the Factors Which Influence Satisfaction with Nanhuan Neighborhood

6.5.1 Identification of Factors

The influencing factors of satisfaction were examined through a structural equation model (SEM), which has proven powerful and consistent by several pieces of research on similar topics related to the built environment [45]. SEM is a statistical technique for evaluating the causal relationship between single indicators, their overall role, and latent variables which cannot be observed directly.

The Kaiser Meyer Olkin and Bartlett tests indicated that the data and information collected were suitable for the factor analysis, so an exploratory factor analysis (EFA) was first applied (Table 6.5). EFA groups the original factors into some influencing factors with strong explanatory ability. Applying the statistical method of maximum variance orthogonal rotation, the influencing factors resulted: "Individual attributes", "Self-conditioned cognition", "Community environment", "Greening level", "living before New Nanhuan Village", "Living condition", "Living life", "Residents' interaction" and "Supporting facility supply"; Table 6.6 lists the influencing factors and what variables they refer to.

6.5.2 Modelling

Starting from the influencing factors condensed by the exploratory factor analysis in Table 6.6, and the hypothesized relationships among the variables, IBM® SPSS® Amos 27, structural equation modelling program, was used to construct the SEM; the statistical method of maximum likelihood estimation was applied to calculate

Table 6.6 Coefficient results of SEM after correction

			Estimate	S.E	C.R	*P*	Estimate(S)
Age	←	Individual attribute	1				1.199
Occupation	←	Individual attribute	−0.419	0.089	−4.69	***	−0.542
Shopping preference	←	Individual attribute	0.311	0.038	8.121	***	0.647
Property management	←	Community environment	1				0.857
Building quality	←	Community environment	0.596	0.068	8.793	***	0.54
Entrance convenience	←	Community environment	0.422	0.091	4.625	***	0.286
Air quality	←	Community environment	0.545	0.099	5.496	***	0.338
Green space landscape	←	Greening level	1				1.117
Current residence	←	Greening level	0.335	0.113	2.975	0.003	0.635
Dwelling area	←	Living condition	1				0.779
Housing type	←	Living condition	0.678	0.089	7.659	***	0.71
Education	←	Residence life	1				0.549
People/each unit	←	Residence life	−4.034	0.787	−5.126	***	−0.983
Pedestrian safety	←	Self-conditioned cognition	1				0.515
Income	←	Self-conditioned cognition	−1.849	0.507	−3.646	***	−0.628
Gender	←	Self-conditioned cognition	0.303	0.069	4.426	***	0.308
Improvement after regeneration	←	Residence change	1				0.141
Household register	←	Residence change	1.719	0.721	2.385	0.017	0.818
Residence before moving	←	Residence change	2.563	1.068	2.401	0.016	0.696
Residents' satisfaction	←	Individual attribute	0.033	0.033	1.005	0.315	0.05
Residents' satisfaction	←	Supporting facility	0.011	0.029	0.387	0.699	0.016
Residents' satisfaction	←	Community environment	0.822	0.084	9.79	***	0.844
Residents' satisfaction	←	Greening level	0.039	0.04	0.988	0.323	0.051

(continued)

Table 6.6 (continued)

			Estimate	S.E	C.R	*P*	Estimate(S)
Residents' satisfaction	←	Residence change	0.043	0.245	0.175	0.861	0.014
Residents' satisfaction	←	Self-conditioned cognition	0.709	0.221	3.217	0.001	0.513
Residents' satisfaction	←	Residence life	0.155	0.191	0.812	0.417	0.058
Residents' satisfaction	←	Residence condition	−0.003	0.085	−0.035	0.972	−0.003
Residents' satisfaction	←	Neighbor interaction	0.049	0.024	2.02	0.043	0.082

***Indicates no significance ($P < 0.001$)

the value of the variables in the model. Among them, "included variables" is the observation variable, and "factor name" is the latent variable, the two are independent variables with the corresponding relationships in the model. "Residents' satisfaction" is the dependent variable. Link the collected model data to the modified model, and correspond each variable to the frame line in the path diagram, then select "name unobserved variables" to name the latent variables and residuals, and finally select "analysis properties" to set the output options, including "optimization history", "standardized estimates", "modification indicators", "indirect, direct & total effects" "Factor score weights", "variances of estimates", "tests for normality and outliers", and finally "calculate estimates" to get the analysis results.

The first model was composed of nine latent (unobserved) variables and 22 explicit ones, including 21 exogenous variables and one endogenous variable. The first model required corrections because coefficients can only be estimated in the just-identified or over-identified model and if most of the observed and latent variables are at the confidence level, while in the estimate results, some paths relationship and regression coefficients did not meet the requirements—the standardized coefficient smaller than 0.1—and were not significant, as an example the path 'Residents' satisfaction ← Individual attributes' (Table 6.6).

The model was modified using the correction index provided by Amos: the samples data of the 22 variables were substituted removing relationship paths that proved not significant and adding influence paths that improved the correlation between model fitness and error variables (Fig. 6.3).

Each observation variable's C.R. and P values were within a reasonable confidence range. Some test indexes were applied to assess the model performance after the correction, such as Comparative Fit Index, Root Mean Square Error of Approximation, Relative Chi Square, and a good fitting degree was found (Table 6.7).

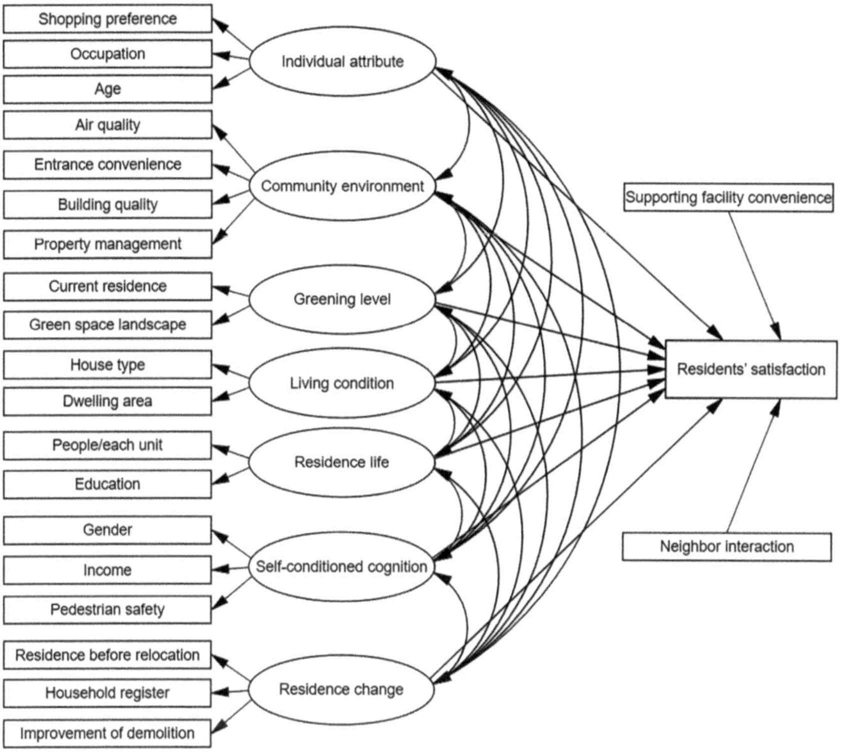

Fig. 6.3 Diagram of SEM after correction

Table 6.7 Results of the overall performance tests

Test index	Ideal standard	Model results	Judgment of model adaptation
CFI—Comparative Fit Index	>0.9	0.997	True
CMIN/DF—Relative Chi-Square	<3.00	1.027	True
GFI—Goodness of Fit Index	>0.8	0.956	True
NFI—Normed Fix Index	>0.9	0.910	True
RMR—Root Mean square Residual	<0.05	0.035	True
RMSEA—Root Mean Square Error of Approximation	<0.05	0.009	True
TLI—Tucker-Lewis Coefficient	>0.9	0.996	True

6.5.3 SEM Analysis Results

As shown in the SEM result of Table 6.7, some indicators resulted to have little corelation to the residents' satisfaction, i.e. $P < 0.001$, they are "Individual attribute" (3 indicators), "Supporting facility" (1 indicator), "Greening level" (2 indicators), "Residence change" (3 indicators), "Residence life" (2 indicators), and "Living Condition" (2 indicators); the reasons of this little significance and the consequent not consideration of the indicators are as follows: (1) Individual attributes "Residence Change" and "Residence Life": even if the community was partially transformed the living conditions of the residents groups in the community are perceived as very similar. We found that the residents in Nanhuan New Village and the residents in the Old Community enjoy similar housing conditions and living characteristics; in addition, we found as well that nearly 60% of the residents in the current community are households that lived in situ before transformation. (1) Indicators "Supporting facility," "Greening level," and "Living Condition": the primary supporting facilities and the greening available to the residents of Nanhuan are in the new part of the community, result of the partial transformation; this way the existing service level in the whole community shows a very small difference. The same considerations can be made to the Living Conditions indicators.

The SEM analysis shows that "community environment", "self-conditioned cognition", and "neighbour interaction" are the most relevant influencing factors of residents' satisfaction with the whole Nanhuan community. In both parts of the neighbourhood, "Community environment" and "self-condition cognition" have a strong correlation with residents' satisfaction (Fig. 6.4), while "neighbour interaction" has little.

Residents pay attention to some factors specifically related to space, such as number of rooms and dwelling area. Still, the results show that property management is the main influencing factor in absolute terms. This finding can be surprising, but it expresses the great importance of the elements which directly impact everyday life, such as the efficiency and the functioning of the buildings and the compounds. For the renewed part of the community, it can be concluded that the relocation is interpreted as an opportunity [64] and the most considered issues in answering the questionnaire are the ones related to the everyday life in the community environment, once the resettlement problems are settled, such as the compensation for what lost, the unit dimension and the housing conditions.

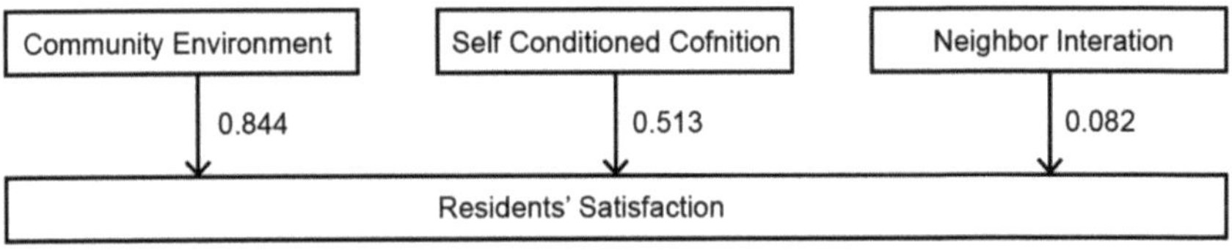

Fig. 6.4 The influencing factors of residents' satisfaction with Nanhuan community

In addition, in the significant relationship, we also found some negative correlation factors, such as "Occupation ← Individual attribute," "People/each unit ← Residence life," and "Income ← Self-conditioned cognition." Negative correlation means the more stable work "occupation", the greater the negative impact on the "Individual Attribute. The more people living in each unit, the greater the negative impact on "Residence life." The higher the "Income", the greater the negative impact on "Self-conditioned cognition".

The findings obtained in Nanhuan are very similar to the ones found in literature about research on residents' satisfaction. Unstable occupation, more people living in the residential unit, the reduced personal living space will decrease the residents' satisfaction [61, 65]. Higher-income people have usually higher demand for living conditions so their satisfaction with the community is low [66], but they have also the economic resources to get what they need or desire outside the community, so what impacts the most their satisfaction is the quality of the residential units.

6.6 Conclusion and Discussion

6.6.1 Discussion

The residents' satisfaction evaluation in Nanhuan shows that most of the residents are either almost satisfied or satisfied with their living conditions, both the ones who live in the part of the community which was not regenerated and the ones who live in the new part. This finding is surprising and allows two different conclusions: on one side the impact of the regeneration which radically changed the characters of the community is modest, and densification is not having a negative impact on people satisfaction, on the contrary it seems irrelevant. On the other side we can say that the regeneration has a very positive impact because it makes available public facilities and everyday shopping opportunities, which before the regeneration were not existing of limited. We can conclude that what matters in this community of low-middle income families is not the residential unit they inhabit, but the urban environment and its public facilities it offers. In fact, the factors "community environment" and "surrounding facilities" are the ones with the highest satisfaction level. On the contrary, "building quality" and "property management" are relatively unsatisfactory, indicating that especially the New Village should pay more attention to community management.

Given these findings, this study concludes that neither the urban form nor the building type, no matter their age and dimension, once the basic conditions are fulfilled, are the main factor of the residents' satisfaction of Nanhuan, but the public facilities and the way the community is managed. This result is especially striking, because of the radical difference between the built environment of the old and the new part of the neighborhood, and because the main reason for this radical transformation was a living condition that was assessed as not satisfactory by the promoters of the initiative.

The differences between the parts, both in the general assessment and in the single factors which influence the satisfaction, do not indicate different orientations of residents' satisfaction, which allows concluding that both urban forms of Nanhuan—the traditional resettlement and the high-rises neighborhood—are perceived as good enough. Densification is not an issue if it does not produce the perception of overcrowding, and it does not in the case studied. This homogeneity of assessment allows concluding that the spatial differences do not significantly impact the assessment, and the two parts, which host populations with similar characteristics, still perceive themselves as one community, even though they occupy completely different urban forms. This can indicate that the regeneration project was successful in the requalification but failed to offer a radically improved urban experience and quality; probably, only a much larger intervention could have achieved this impact.

However, it can as well be concluded that the regeneration of one part of the neighborhood was able to improve also the other part with public facilities, the green landscape, and easily accessible commercial activities; the case study shows how the improvement of a whole site can be achieved by the intensive redevelopment of one limited area of the site.

In addition, it must be recognized that the Nanhuan project improved the conditions without provoking gentrification, and the residents did share the land added value due to the transformation, which is very probably one of the unspoken reasons for the overall satisfaction expressed by the residents. The fact that this regeneration project had no intention to produce gentrification and gentrification did not happen is the main element of its social sustainability. Also, the fact that the residents were engaged in a consultation process during the last phase of the realization of the new Nanhuan is an action towards the social sustainability of the regeneration output.

It must be also noted that some of the transformations of Nanhuan Village—such as well-designed green space and places for the community life, convenient public transport connection, and multiple facilities—matter for improving social sustainability. The results showed that some fundamental human needs such as "air quality" "convenience" and "building quality" affect satisfaction and should be considered in every stage of the regeneration process. The results also show self-conditioned cognition and neighbor interaction are significant factors affecting the sustainability of the community life: a well-organized community management could guarantee orderly daily life. Residents' engagement and empowerment could as well reach the level of promoting the self-upgrading of the spaces and structures.

6.6.2 Conclusion

Residents' satisfaction with the living environment is increasingly considered in China for achieving people-oriented development in a harmonious society, which is a national goal declared by the 13th five-year plan of China; people-oriented development is the basis for socially sustainable development. Urban regeneration is as well declared a national goal by the 14th five-year plan to increase the efficiency of land use and improve the living quality. Given these national policies, this study analyzed the residents' satisfaction with the living environment of their resettlement community, Nanhuan neighborhood in Suzhou, before and after its government-led regeneration.

The reason for conducting this assessment is few studies have compared the residents' satisfaction before and after the transformation of their living conditions and analyzed. The study is relevant also because it focuses on a case of housing reconstruction with increased FAR, and densification can be distinguished in developed China especially in the transformation of old neighborhoods, where old means built before 2000 [15, 67]. This trend is very relevant for the future of urbanization in China and goes along with the increase of market-oriented transformations in the governance of urban development, therefore the analysis of the main factors affecting residents' satisfaction with Nanhuan can be useful for similar projects and for drafting processes of participation with multiple subjects [68, 69].

This study provides a residents' satisfaction evaluation questionnaire framework design reference for reconstructing of old communities. Social sustainability relies on creating a quality of life at the community level, but the specific implementation path and carrier are unclear. By measuring residents' satisfaction as a link, the study contributes to further choices about the transformation of resettlement communities built before 2000 and achieved socially sustainable regeneration [70].

Community regeneration planning is limited by multiple factors such as land use and investment quota, and the planning area and implementation contents are greatly limited. The residents' satisfaction evaluation can directly reflect the influencing factors of different environmental and social factors. The remarkable satisfaction indicators can assist in the formulation of regeneration discission-making. The planning could strengthen the aspects with substantial influencing factors and ensure a significant improvement after the implementation according to the plan.

The study concentrated on the living conditions after the regeneration project to assess if the urban quality was perceived as improved by the inhabitants, and further investigation should focus on the process of transformation, especially on the compensation after relocation issue and people's participation in the definition of the design solution, which can impact substantially the satisfaction level and the design of the site. The study shows how it is necessary to determine the priorities of community planning in combination with field investigation, spatial analysis, and people's involvement to define the most appropriate planning strategies for the site [71].

With the available resources, the research concentrated on the interviews and questionnaires, carried out in the amount, which was what necessary for the modelling, in future we will expand the research integrating the results of the modelling with qualitative research and assessment such as urban syntax analysis.

Acknowledgments The author gratefully acknowledges Yunqing Xu, Geng Ma, Haoqi Wang, Yang An, Yihan Shi, and Xiaoxiao Feng for their contributions to data acquisition and for offering insightful research assistance and comments on earlier versions of the research. Additionally, the figure on page 103 was originally developed by Wenkang Tian (Harvard University).

References

1. State Council of the People's Republic of China (2021) Outline of the 14th Five-Year Plan (2021–2025) for national economic and social development and long-range objectives through the year 2035 [R]. Beijing: Central Party Literature Publishing House. Translated by CCPC Party History and Literature Research Institute. (Official English version)
2. Ji M (2015) Urban housing security satisfaction in China: an HLM model and empirical study based on the 2012 Lian's public service quality of Chinese cities survey. E-GovMent 150:67–75
3. Zhang X, Chen J, Wang H, Yang D (2025) From policy synergy to equitable greenspace: unveiling the multifaceted effects of regional cooperation upon urban greenspace exposure inequality in China's megacity-regions. Appl Geogr 174:103472. https://doi.org/10.1016/j.apgeog.2024.103472
4. Chen J, Li P, Wang H (2025) Assessment of influential factors on commute and life satisfaction in a historic campus-adjacent environment: evidence from a comparison study of Twin Cities. J Urban Plan Dev 151:4024065. https://doi.org/10.1061/JUPDDM.UPENG-5171
5. Chen J, Ren K, Li P, Wang H, Zhou P (2024) Toward effective urban regeneration post-COVID-19: urban vitality assessment to evaluate people preferences and place settings integrating LBSNs and POI. Environ Dev Sustain. https://doi.org/10.1007/s10668-024-05378-8
6. Akbar PNG, Edelenbos J (202) Positioning place-making as a social process: a systematic literature review. Cogent Soc Sci 7
7. Mensah J (2019) Sustainable development: meaning, history, principles, pillars, and implications for human action: literature review. Cogent Soc Sci 5
8. Barton H (2000) Sustainable communities: the potential for eco-neighbourhoods. Earthscan, ISBN 1853835137
9. Tzoulas K, Korpela K, Venn S, Yli-Pelkonen V, Kaźmierczak A, Niemela J, James P (2007) Promoting ecosystem and human health in urban areas using green infrastructure: a literature review. Landsc Urban Plan 81:167–178. https://doi.org/10.1016/j.landurbplan.2007.02.001
10. Zhao X, Chen J, Li J, Wang H, Zhang X, Yu F (2025) Unraveling the renewal priority of urban heritage communities via macro-micro dimensional assessment-a case study of Nanjing city, China. Sustain Cities Soc 124:106317
11. Song H, Chen J, Li P (2025) Decoding the cultural heritage tourism landscape and visitor crowding behavior from the multidimensional embodied perspective: Insights from Chinese classical gardens. Tour Manag 110:105180. https://doi.org/10.1016/j.tourman.2025.105180
12. Pan L, Gan W, Chen J, Xian X (2024) Ecological planning for mitigating urban heat based on thermal network simulation: a case study of Guangzhou-Foshan metropolitan area. J Chin Urban For 22:10–19. https://doi.org/10.12169/zgcsly.2024.06.26.0004
13. Guan X, Wei H, Lu S, Dai Q, Su H (2018) Assessment on the urbanization strategy in China: achievements challenges and reflections. Habitat Int 71:97–109. https://doi.org/10.1016/j.habitatint.2017.11.009
14. Chu Y (2020) Wah China's new urbanization plan: progress and structural constraints. Cities 103:102736
15. Pellegrini P, Chen J (2020) Hypothesis of densification for a sustainable urbanization in a wealthy Chinese city. IOP Conf Ser Earth Environ Sci 588
16. Chen J, Zhao X, Wang H, Yan J, Yang D, Xie K (2024) Portraying heritage corridor dynamics and cultivating conservation strategies based on environment spatial model: an integration of multi-source data and image semantic segmentation. Herit Sci 12:419. https://doi.org/10.1186/s40494-024-01497-7

17. Chen J, Li P, Wang H (2024) MSPA-based collaborative evaluation and planning governance of territory and natural environment in metropolitan agglomerations—a case study of EGIDDZ in Yangtze River Delta. J Resour Ecol 15:1263–1273. https://doi.org/10.5814/j.issn.1674-764x.2024.05.014
18. Yang N, Chen J, Ban L, Li P, Wang H (2024) Pre-planning and post-evaluation approaches to sustainable vernacular architectural practice: a research-by-design study to building renovation in Shangri-La's Shanpian House, China. Sustain 16:9568
19. Chen J, Wang H, Yang Z, Li P, Ma G, Zhao X (2023) Comparative Spatial vitality evaluation of traditional settlements based on SUF: taking Anren ancient town's urban design as an example. Sustain 15:8178
20. Li H, Liu Z (2019) Analysis on the spatiotemporal evolution of urban density and high-density development in China—from 1981 to 2014. Urban Dev Stud 26:9
21. State Council of the People's Republic of China. Regulations on expropriation of buildings on state-owned land and Compensation [s] (State Council Decree No. 590). Beijing: State Council, 2011-01-21. http://www.gov.cn/zhengce/content/2011-01/21/content_2617838.htm. Accessed 30 June 2025
22. Qian Z (2019) Displaced villagers' adaptation in concentrated resettlement community: a case study of Nanjing, China. Land Use Policy 88:104097
23. Ma G, Paola P, Wu H, Han H, Wang D, Chen J (2023) Impact of land-use mixing on the vitality of urban parks: evidence from big data analysis in Suzhou, Yangtze River Delta region, China. J Urban Plan Dev 149:4023045. https://doi.org/10.1061/JUPDDM.UPENG-4334
24. Chen J, Tian W, Xu K, Pellegrini P (2022) Testing small-scale vitality measurement based on 5D model assessment with multi-source data: a resettlement community case in Suzhou. ISPRS Int J Geo-Inf 11:626. https://doi.org/10.3390/ijgi11120626
25. Chen J, Pellegrini P, Wang H (2022) Comparative residents' satisfaction evaluation for socially sustainable regeneration—the case of two high-density communities in Suzhou. Land 11:1483. https://doi.org/10.3390/land11091483
26. Xie Y, Hu J (2014) An introduction to the China Family Panel Studies (CFPS). Chin Sociol Rev 47:3–29
27. Chen J, Pellegrini P, Ma G (2021)Identifying resettlement communities' urban regeneration opportunity through GIS-based spatial analysis in Suzhou metropolitan area. Urban Reg Plan 6:146. https://doi.org/10.11648/j.urp.20210604.15
28. Zhang M, Qiao S, Yeh AGO (2021) Blemish of place: territorial stigmatization and the depreciation of displaced villagers' resettlement houses in Chengdu, China. Cities 117:103330
29. Xi J (2021) speech at the celebration of the 100th anniversary of the founding of the communist party of China. Ind China 8–19
30. Dempsey N, Bramley G, Power S, Brown C (2011) The social dimension of sustainable development: defining urban social sustainability. Sustain Dev 19:289–300. https://doi.org/10.1002/sd.417
31. Karuppannan S, Sivam A (2011) Social sustainability and neighbourhood design: an investigation of residents' satisfaction in Delhi. Local Environ 16:849–870. https://doi.org/10.1080/13549839.2011.607159
32. Chen J, Li P, Lei Y, Zhang Y, Lai C, Chen B, Liu J, Schnabel MA (2025) Leveraging augmented reality for historic streetscape regeneration decisionmaking: a big and small data approach with social media and stakeholder participation integration. Cities. https://doi.org/10.1016/j.cities.2025.106214
33. Diener E, Ryan K (2009) Subjective well-being: a general overview. S Afr J Psychol 39:391–406
34. Mohit MA (2014) Present trends and future directions of quality-of-life. Procedia—Soc Behav Sci 153:655–665
35. Huang X, He D, Tang S, Li X (2020) Compensation, housing situation and residents' satisfaction with the outcome of forced relocation: evidence from urban China. Cities 96:102436. https://doi.org/10.1016/j.cities.2019.102436
36. The State Council several opinions of the State Council on deepening the construction of new urbanization

37. Chen F, Chen H, Zhu Z, Peng B (2000) The evaluation of urban residential quality and satisfaction. Hum Geogr 25–27
38. Ma L, Woods O, Zhu H (2019) Restoration of an ancestral temple in Guangzhou, China: re-imagining history and traditions through devotion to art and creation. Cult Geogr 26:141–149
39. Li Z, Wu F (2013) Residential satisfaction in China's informal settlements: a case study of Beijing, Shanghai, and Guangzhou. Urban Geogr 34:923–949
40. Geng Y (1999) Evaluation and method of residential satisfaction. J Tsinghua Univ 14:79–85
41. Chen X (2005) Public participation mechanism and Green Community creation. Straits Sci 000:1–3
42. State Council of the People's Republic of China. Action plan for green community creation in urban and rural construction [R] (Policy document No.: Jian ban Cheng [2020] No. 68). Beijing: Ministry of Housing and Urban-Rural Development, 2020-07-15. http://www.mohurd.gov.cn. Accessed 30 June 2025
43. Li Y, Zhang X, Xia C (2023) Towards a greening city: how does regional cooperation promote urban green space in the Guangdong-Hong Kong-Macau Greater Bay Area? Urban For & Urban Green 86:128033
44. Zhang X, Shen J, Sun Y, Zhou C, Yang Y (2023) Formation of city regions from bottom-up initiatives: investigating coalitional developmentalism in the Pearl River Delta. Ann Am Assoc Geogr 113:700–716
45. Yuan Y, Ding K, Cao X, Wu X (2018) A review of neighborhood satisfaction. Urban Dev Stud 25:111–117
46. Huang Z, Du X (2015) Assessment and determinants of residential satisfaction with public housing in Hangzhou, China. Habitat Int 47:218–230
47. Tan Y, He J, Han H, Zhang W (2019) Evaluating residents' satisfaction with market-oriented urban village transformation: a case study of Yangji village in Guangzhou, China. Cities 95:102394. https://doi.org/10.1016/j.cities.2019.102394
48. Niu J, Zhao M (2018) Analysis on factors influencing of yurts residence satisfaction based on structural equation model. Sci Dev 14:750–756
49. Yi L, Jianping L (2014) The Influence factors of residential satisfaction of urban floating population in China. Chongqing Soc Sci 61–68
50. Wang X, Zhang J, Li L (2014) A survey of college graduates' housing satisfaction—a case study of Jiaxing talent apartment. Urban Probl 95–101
51. Dang Y, Yu J, Zhang W, Li Y, Chen L, Zhan D (2016) Satisfaction evaluation of living environment and influencing factors in the Bohai Rim area. Prog Geogr 35:184–194
52. Riazi M, Emami A (2018) Residential satisfaction in affordable housing: a mixed method study. Cities 82:1–9. https://doi.org/10.1016/j.cities.2018.04.013
53. Wang D, He S, Webster C, Zhang X (2019) Unravelling residential satisfaction and relocation intention in three urban neighborhood types in Guangzhou, China. Habitat Int 85:53–62
54. Lee KY (2022) Relationship between physical environment satisfaction, neighborhood satisfaction, and quality of life in Gyeonggi, Korea. Earth Its Atmos 10:663
55. Wang P, Qin X, Li Y (2021) Satisfaction evaluation of rural human settlements in northwest China: method and application. Land 10:813
56. Zhang X, Chen Y, Xu Y, Yue W, Tang C (2023) Does regional cooperation constrain urban sprawl? Evidence from the Guangdong-Hong Kong-Macao Greater Bay Area. Landsc Urban Plan 235:104742
57. Zhang X, Li Y, Zhou C, Luan X, Yuan F (2023) Rescaling of the land regime in the making of city-regions: a case study of China's Pearl River Delta. Urban Stud 60:483–500
58. Yuan Y, Yang M, Gan Z, Wu J, Xu C, Lei D (2019) Analysis of the risk factors affecting the size of fatal accidents involving trucks based on the structural equation model. Transp Res Rec 2673:112–124
59. Cao XJ (2016) How does neighborhood design affect life satisfaction? Evidence from Twin Cities. Travel Behav Soc 5:68–76. https://doi.org/10.1016/j.tbs.2015.07.001
60. Cao J, Zhang J (2016) Built environment, mobility, and quality of life. Travel Behav Soc 5:1–4

61. Mohit MA, Azim M (2012) Assessment of residential satisfaction with public housing in Hulhumale', Maldives. Procedia—Soc Behav Sci 50:756–770
62. Zou J, He Z, Zhai G, Hu Y (2018) Determinants of commuting distance based on structural equation model in case of Nanjing. Mod Urban Res 10:80–108
63. Zou H, Luo X, Tu J (2014) An empirical study on living satisfaction of the restricted property house: a case study of Maigaoqiao in Nanjing. Hum Geogr 29:61–65
64. Kearns A, Mason P (2013) Defining and measuring displacement: is relocation from restructured neighbourhoods always unwelcome and disruptive. Hous Stud 28:177–204
65. Parkes A, Kearns A, Atkinson R (2002) What makes people dissatisfied with their neighbourhoods? Urban Stud 39:2413–2438
66. Chan E, Lee GKL (2008) Critical factors for improving social sustainability of urban renewal projects. Soc Indic Res 85:243–256. https://doi.org/10.1007/s11205-007-9089-3
67. Pellegrini P, Chen J (2021) Sustainable urban renewal and densification in China: the case of Suzhou in the Yangtze River Delta region
68. Lin Y, De Meulder B (2012) A conceptual framework for the strategic urban project approach for the sustainable redevelopment of "villages in the city" in Guangzhou. Habitat Int 36:380–387
69. Gao J, Chen W, Liu Y (2018) Spatial restructuring and the logic of industrial land redevelopment in urban China: II. A case study of the redevelopment of a local state-owned enterprise in Nanjing. Land Use Policy 72:372–380
70. Zhao X, Yu F, Zhang X, Chen J, Li P (2025) Assessing Urban Renewal Efficiency via Multi-Source Data and DID-Based Comparison between Historical Districts. npj Herit Sci 13:389. https://doi.org/10.1038/s40494-025-01964-9.
71. Pellegrini, Paola; Chen, Jinliu The Regeneration Path Not Taken Further: An Experiment in Urban Densification and State Entrepreneurialism in a Resettlement Neighborhood in Suzhou, China. Urban Stud. 2025, 00420980251365188, doi: https://doi.org/10.1177/00420980251365188.

Chapter 7
Comparative Residents' Satisfaction Evaluation for Social Sustainable Regeneration: The Case of Two High-Density Communities in Suzhou

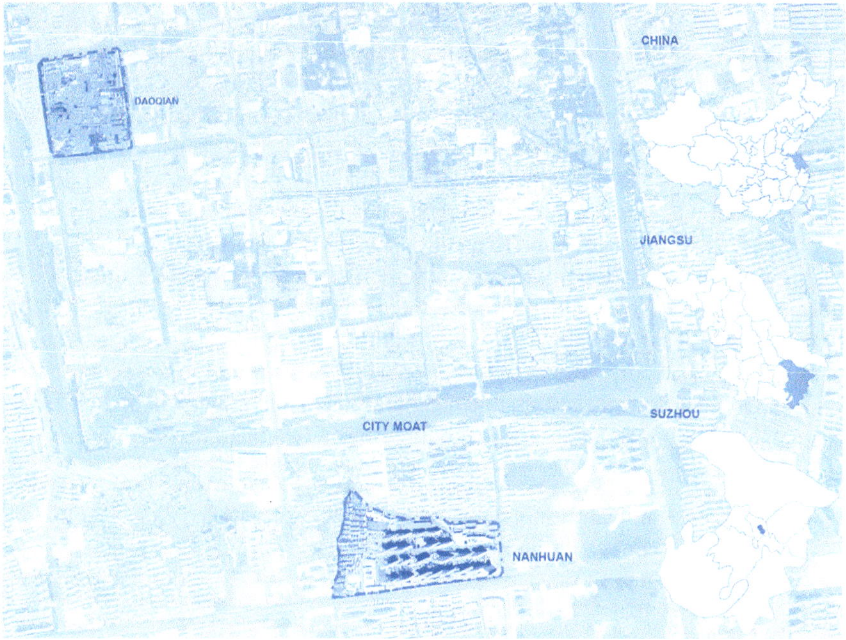

7.1 Introduction

Since China's 13th national Five-Year Plan, social sustainability has been highlighted as an essential component for the realization of high-quality urbanization [1]. With the urbanization rate in 2021 reaching 64.7%, an urban regeneration approach to development was first mentioned in the Chinese 14th five-year plan (2021–2026).

J. Chen and P. Pellegrini, *Densification as a Design Tool for Sustainable Urban Regeneration*, Urban Sustainability, https://doi.org/10.1007/978-981-95-1653-7_7

Combined with the idea of people-oriented development, regeneration aims to optimize urban spatial structures and improve urban quality [2, 3]. The regeneration of old and obsolete residential communities, as the basic units of social space, is a critical step to achieving high-quality urbanization [4, 5]. The regeneration movement pays attention to residents' sense of happiness, improves residents' satisfaction, and finally, creates socially sustainable communities [6].

The regeneration of old communities is a controversial process. In fact, the community regeneration process can easily cause social problems such as segregation, gentrification and inequality [7]. Therefore, social sustainability is particularly important when evaluating the effectiveness of old community regeneration projects. A socially sustainable community can be defined as a place where people want to live and work in the long term [8], and where the enhanced living environment meets residents' needs and ensures social justice [9].

Improving quality of life (QoL) is vital for enhancing social sustainability [10]. On one hand, the living environment can significantly affect the QoL. The measurement of QoL includes objective indicators of the built environment, which can assist in fulfilling living requirements. On the other hand, under the guidance of the current policy, the transformation process needs to focus more on residents' satisfaction [11]. Measurements of QoL include comprehensive consideration of subjective feelings, personal well-being, social balance and social justice [12].

Therefore, measuring resident satisfaction is a very effective way to achieve improved QoL, to evaluate the sustainable development of old communities, to measure infrastructure status and to achieve social sustainability [13, 14].

China's ongoing urban regeneration initiative aims to improve the QoL by improving the living environment. Nevertheless, although QoL was recently highlighted as a national goal, there is still very limited research on how to regenerate old communities. Therefore, it is necessary to study how to combine resident satisfaction with QoL and socially sustainable regeneration. In the framework of the Chinese urban regeneration movement, this study focuses on two main research questions:

1. How to build a residents' satisfaction evaluation system based on the QoL index?
2. What factors significantly affect residents' satisfaction in two typical but radically different residential communities?

7.2 Conceptual Framework

Under China's current urban regeneration movement, this section illustrates the correlation between social sustainability, quality of life, and old community regeneration. This section also introduces how social sustainability could be applied at the community level and how QoL could help form residents' satisfaction evaluation indicators.

Table 7.1 Quality of life evaluation index for social sustainability community regeneration

Housing indicators	Housing ownership, housing area, per capita housing area
Neighborhood index	Educational facilities, business services, entertainment services, transportation services, open space, safety, landscape, social environment (community), visual perception
Socioeconomic	Age, gender, income, education level

7.2.1 Social Sustainability and QoL

As social sustainability is a broad concept, the term has a range of meanings. In this chapter, it describes the social goals of sustainable development [15]. In many explanations, social sustainability includes social capital, social equity, and public participation. At the community level, social sustainability implies a sustainable and positive QoL based on an understanding people's needs and sense of belonging [16].

QoL is a comprehensive synthesis of individual well-being and social balance, objective indicators and subjective feelings [17]. In the narrow sense, QoL quantifies psychological acceptance, individual characteristics, cognition, etc. on the part of the individual. In the broad sense, it encompasses social equity, including degrees of life satisfaction of different people and future generations (i.e., according to the objective material environment). The relationship between QoL and social sustainability at the community level is predicated upon three dimensions: housing indicators, neighborhood index, and socioeconomic indicators (Table 7.1):

1. Housing indicators: housing area, building quality, house type structure, ventilation, and lighting [18].
2. Neighborhood indicators: facilities in the community, such as shops, educational facilities, and public transport [19]. Elderly care service measurements evaluate the level of elderly care and medical and health care[20], as well as public spaces [21].
3. Socioeconomic indicators: individual characteristics and perception of QoL, such as gender, income level, home ownership [22].

7.2.2 People-Oriented Old Community Regeneration and Social Sustainability

Current approaches to the regeneration of old communities have the following characteristics:

1. Government-oriented development in response to the renovation of old communities, with major implementation strategies focusing on "wearing clothes and hats", that is, beautifying buildings and upgrading infrastructure.
2. For different types of old communities, the applied approaches are the same.

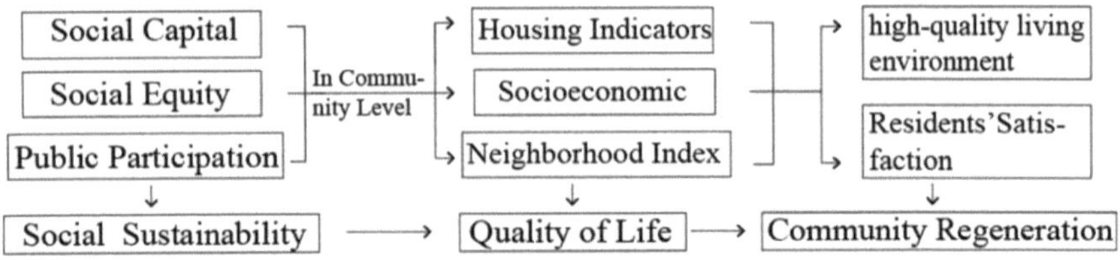

Fig. 7.1 QoL as the mediation link social sustainability and old community regeneration. (*Source* Drawn by the author)

3. There are many contradictions in the regeneration process. Community planners focus on collecting and coordinating the willingness of residents to accept regeneration initiatives.

The people-oriented regeneration mentioned in the 14th national Five-Year Plan is intended to promote a sense of happiness and well-being for residents, from physical space, standardization, and implementation to social, incentive, and comprehensive regeneration [23]. The interest among old communities in a high-quality living environment and resident satisfaction were informed by assessments of QoL [15, 24–26]. In this way, through the prism of QoL, social sustainability and community regeneration could be synchronized (Fig. 7.1).

7.2.3 QoL and Residents' Satisfaction Evaluation

Residents' satisfaction usually refers to the satisfaction of residents living in a specific place [27]. It is a tool that reflects opinions about the housing status quo, planning proposals, and policy making [28]. Satisfaction-influenced factors are widely distributed and determined by the gap between actual and expected living environments [29]. Measurements of satisfaction can reflect residents' QoL. Such assessments must take into account the contradiction between the growing needs of people for a better life and unbalanced and insufficient development [30].

Based on a QoL evaluation index (Table 7.1) of urban regeneration, data regarding residents' life satisfaction levels can be divided into the objective environmental characteristics of residential areas [31] and the subjective feelings of residents [32]. Subjective feelings including individual characteristics, influenced by gender, age, and education level, are also called "internal causes". Objective environmental characteristics ("external factors") are measured at two levels:

1. Built Environment, the impact of which on satisfaction is obvious [33], e.g., an over-crowded, polluted living environment has a negative effect on life satisfaction [34]. Individuals living in poor environments may offset the benefits of life satisfaction to community residents [35].
2. Social Connection, i.e., social ties among community members. For example, greeting and chatting may significantly improve residents' satisfaction [36, 37].

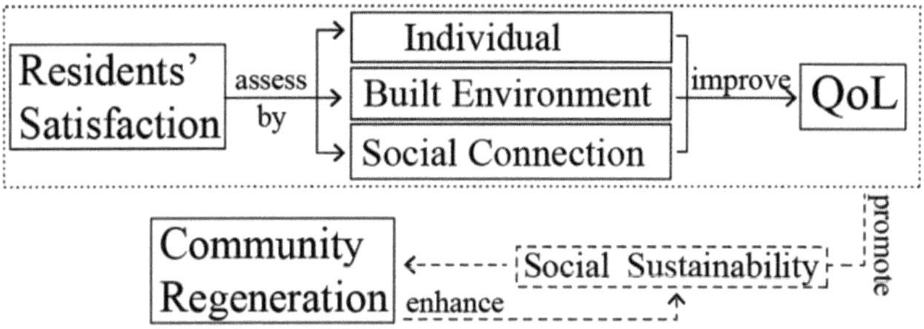

Fig. 7.2 Residents' satisfaction evaluation for social sustainability community regeneration. (*Source* Drawn by the author)

The shopping process accompanied by more or less communication will also affect satisfaction level [38].

Taking into account individual, environment, and social connections when assessing residents' satisfaction is essential for the development of renovation approaches which improve QoL, as well as guiding community regeneration toward social sustainability (Fig. 7.2).

7.3 Data Collection and Methodology

7.3.1 Case Selection

Suzhou was selected as the case study for several reasons:

1. It is one of the main historical and cultural cities in China and is comparatively affluent (its GDP in 2021 was USD $352.22 billion);
2. It ranks 75th in the world and 1st in China in terms of "livability" (Global livable index report). It ranks 58th in the world and 6th in China on the sustainable city index (Kamiya & Ni, 2020);
3. It was chosen by the national government as one of the first pilot cities for urban regeneration, and so is particularly relevant for an examination of sustainable urban regeneration.

Within Suzhou, two residential communities, with different urban morphology and building typology conditions, were selected to compare resident satisfaction; these communities are representative of typical neighborhoods which are replicated throughout the city. They have different spatial features: Daoqian is in the old town and comprises high-density, low-rise, and compact small blocks covering an area of about 18.12 ha, whereas Nanhuan is a resettlement community, made of high-rises with public green areas covering an area of 21.54 ha (Figs. 7.3, 7.4, and 7.5). The

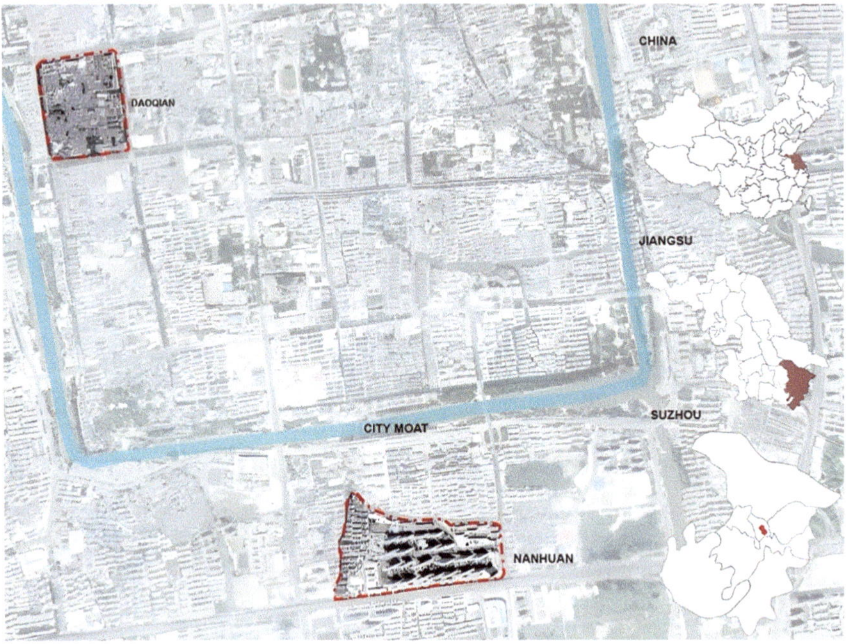

Fig. 7.3 Location of Nanhuan and Daoqian. (*Source* Drawn by the author)

Floor–Area Ratio (FAR) of Nanhuan is 2.58, while that of Daoqian FAR is 1.4. The greening rate of Nanhuan is 25.6% and that of Daoqian is 1.51%.

Nanhuan is located south of Suzhou's ancient city (Fig. 7.3). It comprises a sequence of multi-story buildings constructed for the resettled farmers between the end of the 1970s and the beginning of the 1980s [39, 40]. Due to poor building quality, part of it was demolished and rebuilt into high-rise residential buildings starting in 2010 (Table 7.2). Nanhuan New Village was promoted and realized 10 years earlier than the current national regeneration guidelines and attracted attention due to its radical transformation and the densification of the pre-existing community:

1. It is the first regeneration of a resettlement area to have been planned, funded, and realized by the Suzhou local government. The project was included in the government's annual list of important tasks to set an example for communities experiencing similar conditions.
2. The initiative resulted in the densification and relocation of residents.
3. It combines high-rise buildings with small compact blocks and mixed-use buildings, which is an unusual solution when single function super blocks are most common (Fig. 7.4).

Daoqian community is located in the ancient city (Fig. 7.3). It is a low-rise, high-density residential area with an overall historical style (Table 7.2) and is a good example of the traditional urban morphology in the ancient city of Suzhou:

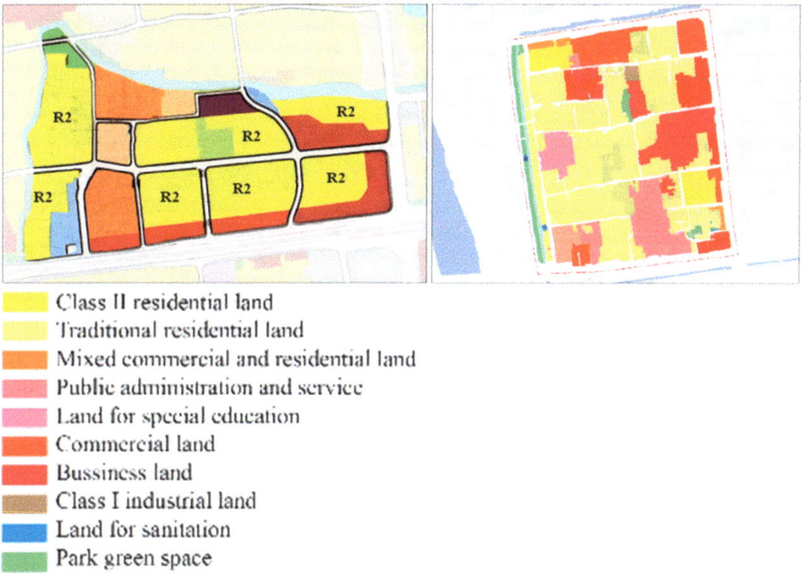

Fig. 7.4 Land use of Nanhuan (left) and Daoqian (right). (*Source* Suzhou local government)

Fig. 7.5 Aerial view of Nanhuan (left) and Daoqian (Right). (*Source* Photographed by the author)

Table 7.2 Existing community construction index

	Nanhuan	Daoqian
Total land area	21.54 ha	18.12 ha
Gross Floor Area (GFA)	555,375.76 m^2	252,450 m^2
Residential area and percentage	391,425 m^2 (70%)	164,092 m^2 (65%)
Floor Area Ratio (FAR)	2.58	1.4
Building density	26.80%	48.84%
Greenery rate	25.60%	1.51%
Units	4852	–
Parking lots	2526	–

1. The community is crisscrossed with alleys and has numerous original buildings which are protected for their historic, cultural, and architectural value;
2. The current land ownership situation is complex, with a mix of individual properties and socially owned ones (Fig. 7.4);
3. The population density in the old town is several times that of other new districts in Suzhou. The green space rate is only 1.5% and the population is aging. Finally, the buildings are old and provide a low-quality living environment [41, 42].

An analysis of the main spatial characteristics of the two communities showed how variables such as FAR, plot ratio, and green space ratio differed in quantity and quality (Table 7.2).

7.3.2 Questionnaire Design and Data Collection

In China, initial studies to measure residents' satisfaction were undertaken relatively late, and various approaches coexisted: some used quantitative and statistical analysis methods to define the influencing factors, while others used qualitative comparisons and descriptive analyses [43, 44].

According to the literature, a comprehensive multi-level index system can be used to analyze residents' satisfaction with their communities. The index system considers objective and subjective factors. Objective factors include the characteristics of the residential unit, the surrounding environment, and infrastructure [45], while subjective factors include personal and family characteristics, income level, house ownership, and compensation for relocation [46, 47].

These factors can be grouped into three main categories:

1. Individual attributes, such as age, education, family structure, economic level, and house ownership [48];
2. Housing conditions, such as building quality, building area, building age, building orientation, lighting, and ventilation [49];
3. Community Context, such as community management, supporting infrastructure, transportation convenience, surrounding environmental conditions, relationships with residents [50].

In addition, as interactions within a community significantly impact its residents' perception, we introduced the category of "intercommunication" [51].

The questionnaire summarized in Table 7.3 comprises five groups of variables to represent a range of the factors and the related degree of satisfaction. The data were recorded using a scale from 1 to 5.

A preliminary field survey was conducted to test the questions and the structure of the questionnaire. In addition, people with different professional knowledge and involvement in the community were interviewed to gain an in-depth understanding of the current situation and to improve the accuracy of data [52, 53].

Table 7.3 Questionnaire structure and content

Satisfaction index	Observational variable
Personal characteristics	Gender
	Age
	Income
	Marital status
	Educational background
Inhabited environment	Area
	Number of members
	Building quality
Community environment	Air quality
	Estate management
	Pedestrian safety
	Green landscape
	People and cars
	Cultural symbols
Community convenience	Public space
	Leisure space
	Parking convenience
	Facility convenience
	Facility diversity
	Ageing services
	Traffic convenience
Intercommunication	Neighborhood interaction
	Shopping tendency
	Overall satisfaction

From August to October 2021, the questionnaire was distributed in both communities by convenience sampling, which is a non-probabilistic sampling method whereby respondents are selected randomly at a specific time and in a specific community area. To this end, 680 questionnaires were distributed, i.e., 330 in Nanhuan New Village and 350 in Daoqian community. The samples are representative because the statistics show how the residents are homogenous in terms of their demographic and socio-economic characteristics [54, 55].

7.3.3 Data and Model Analysis

The first conclusions to be drawn from the results of the questionnaire (Table 7.4) were as follows:

Table 7.4 Questionnaire analysis form of Nanhuan and Danqian community

	Mean value		Standard deviation	
	Nanhuan	Daoqian	Nanhuan	Daoqian
Gender	1.51	1.45	0.501	0.498
Age	1.77	1.02	0.823	0.926
Income	2.42	3.26	1.109	1.473
Area	3.11	2.40	1.233	0.991
Number of members	2.76	2.62	1.213	1.037
Building quality	3.14	3.27	0.788	0.794
Air quality	3.49	3.18	0.841	1.132
Estate management	2.96	3.09	0.982	0.838
Pedestrian safety	3.44	2.99	1.071	0.961
Green landscape	3.02	3.36	1.068	0.814
Public space	2.98	3.63	0.983	0.946
Leisure space	2.87	2.69	0.875	0.723
Parking convenience	2.75	2.35	0.981	1.226
Facility convenience	3.34	3.42	1.037	1.055
Ageing services	3.25	1.63	1.106	0.954
Traffic convenience	3.73	3.63	1.007	0.950
Neighbourhood interaction	3.18	1.68	1.232	1.185
Shopping tendency	1.49	1.62	0.553	0.510
Overall satisfaction	3.26	3.40	0.831	0.692

1. Income: 36.4% of residents in Daoqian community have an income below 5000, while 25.2% of residents in Nanhuan New Village had an income of 10,000–15,000.

   Living area: 64% of residents live in apartments of 50–80 m^2 in size in Nanhuan.
2. The living areas of residents in Daoqian were generally smaller, as would be expected for the houses in the old town (no accurate official data are available).
3. Overall satisfaction: In Nanhuan New Village, scores of 3 and 4 out of 5 were reported by 45.2% and 45.5% of respondents, respectively. The satisfaction scores of Daoqian community were mainly 2, 3, and 4, accounting for 14.2%, 47.3%, and 30.9%, respectively.

In a general overview of the average level of overall satisfaction of the residents, Daoqian, where traditional structures and urban open spaces are conserved, offers worse living conditions, e.g., smaller living areas, to an aging population with a low income than the newer Nanhuan. On account of this, the overall satisfaction was low and the distribution was concentrated (Fig. 7.6). A further detailed analysis showed how the two communities have different expectations.

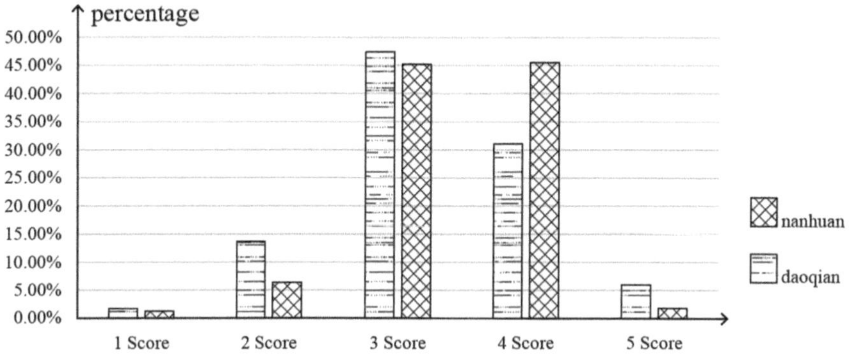

Fig. 7.6 Frequency distribution of satisfaction

7.3.4 *Method Selection*

The SEM is a research method based on statistical analysis technology. It can deal with complex multivariable research data analyses. Scholars proposed a multivariate statistical analysis method to analyze the complex relationship structure between multiple index variables by using a SEM, one of the three significant advances in statistics in recent years [56]. The SEM overcomes the limitations of traditional statistical methods, making it an important tool for multivariate data analyses. It is suitable for three-dimensional and multi-level analyses and can exist in human thinking forms. It can analyze variables (i.e., latent variables) that cannot be directly measured, quantify the causal relationships among various factors, and carry out various subdivisions and comparisons. As shown in Fig. 7.7, A and B are observation variables used to characterize latent variable C, and C influences H. Therefore, this chapter used SEM to analyze the results from our resident satisfaction questionnaire [57, 58].

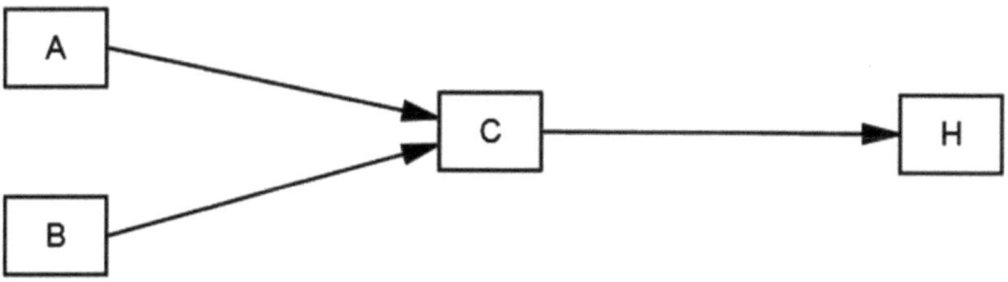

Fig. 7.7 SEM analysis method

7.4 Results and Discussions

7.4.1 Identification of Factors and Modelling

In SPSS, we applied the Kaiser Meyer Olkin and Bartlett tests to ensure that the results were in the normal range and, therefore, valid. Then, exploratory factor analysis (EFA) was applied. Using the maximum variance method, EFA summarized the original data into several groups of explanatory elements by orthogonal rotation. These elements were called "shallow variables" in the structural program model, and the SEM was constructed on this basis (Tables 7.5 and 7.6).

Using the maximum variance orthogonal rotation statistical method, the influencing factors in Nanhuan were found to be: "surroundings", "socializing", "community", and "character". Table 7.5 shows the significant factors and the corresponding variables. The influencing factors obtained from Daoqian were: "individual", "recreation", "management", and "vitality". Table 7.6 shows these significant factors and the corresponding variables.

Starting from the influencing factors condensed by the exploratory factor analysis and the hypothesized relationships among the variables, IBM® SPSS® Amos 26, a structural equation modelling program, was used to construct the SEM. The statistical method of maximum likelihood estimation was applied to calculate the value of the variables in the model (see Fig. 7.8).

Table 7.5 Exploratory factor analysis of Nanhuan

Factor name	Included variables (factor loading)
Surrounding	Facility convenience (0.533), estate management (0.784), building quality (0.693), air quality (0.566)
Socialize	Shopping tendency (0.841), age (0.852), neighborhood interaction (0.476)
Community	Traffic convenience (0.818), public space (0.730), leisure space (0.690)
Character	Pedestrian safety (0.688), income (−0.775), gender (0.578)

Table 7.6 Exploratory factor analysis of Daoqian

Factor name	Included variables (factor loading)
Individual	Marital status (0.643), age (0.802), educational background (−0.809)
Recreation	Green landscape (0.685), public space (0.744), leisure space (0.739)
Management	Air quality (0.560), estate management (0.638), pedestrian safety (0.822)
Vitality	Facility diversity (0.628), people and cars (0.784), cultural symbols (0.723)

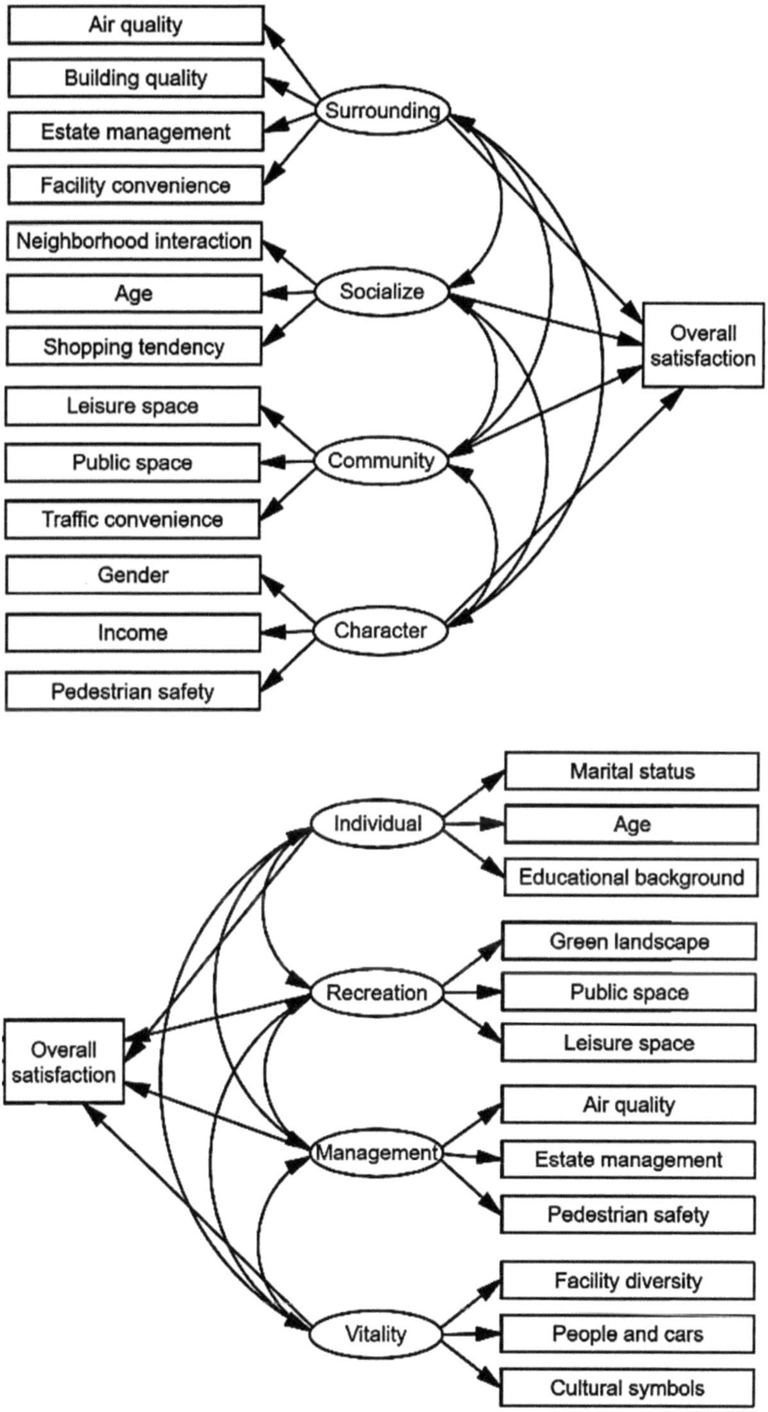

Fig. 7.8 Model structure (Nanhuan-up, Daoqian-down)

7.4.2 SEM Analysis Results

Based on the test results of the structural equation model, and using the correction index provided by AMOS to correct the model, we observed that the "model fit" was within a reasonable range (Table 7.7). We then applied the coefficient results, selected the variables of C.R. > 2, $p < 0.01$, and finally, determined that the degree of life satisfaction of the residents of Nanhuan New Village was determined by the parameters "surrounding", "socialize", "community" and "character" (Table 7.8), of which "surrounding" had the most significant impact, followed by "character". Meanwhile, the degree of life satisfaction experienced by the Daoqian community was determined by "recreation", and "management" (Table 7.9). For the observation variables, the significant factors were "property management", "age", "shopping tendency", and "construction quality".

In the Nanhuan community, "age" and "estate management" had a strong correlation with resident satisfaction, while "facility convenience", "air quality", and "gender" had low correlations. Residents also focused on some factors which are not related to space, such as "community management" and "shopping". Still, the results showed that "estate management" was the main influencing factor, based on the absolute values. This finding may be surprising, but it shows the importance of factors that directly affect daily life, such as the efficiency and function of buildings and complexes. For the renewal part of the community, it could be concluded that resettlement is interpreted as an opportunity [59].

In Nanhuan, "Supporting facilities", "air quality", and "pedestrian safety" had little impact on overall satisfaction (Fig. 7.9). It is worth mentioning that although facilities and green spaces play an important role in people's lives, their impact on satisfaction was not significant due to the high level of renovation and reconstruction that took place in 2014. Public facilities were created between the two parts of the Nanhuan community, and are equally available for both territories. The green space along the canal and the small park in the center of the new part easily accessible to residents of both territories.

Table 7.7 Test results of Nanhuan New Village and Daoqian community

Match index	Reference value	Model result (Nanhuan)	Model result (Daoqian)	Whether it met
CMIN/DF	<3.00	1.239	1.291	Yes
RMSEA	<0.05	0.026	0.03	Yes
RMR	<0.08	0.045	0.038	Yes
NFI	>0.9	0.920	0.906	Yes
TLI	>0.9	0.976	0.965	Yes
CFI	>0.9	0.983	0.976	Yes
GFI	>0.8	0.972	0.968	Yes

CMIN/DF refer to relative chi square

Table 7.8 Modified SEM coefficient results of Nanhuan New Village

			Estimate	S.E	C.R	P	Estimate(S)
Facility convenience	←	Surrounding	1.000				0.324
Estate management	←	Surrounding	2.023	0.377	5.369	***	0.825
Building quality	←	Surrounding	1.355	0.268	5.050	***	0.583
Air quality	←	Surrounding	1.129	0.275	4.113	***	0.340
Shopping tendency	←	Socialize	1.000				0.604
Age	←	Socialize	2.549	0.374	6.826	***	0.846
Neighborhood interaction	←	Socialize	1.883	0.257	7.315	***	0.486
Pedestrian safety	←	Character	1.000				0.494
Income	←	Character	−2.034	0.459	−4.433	***	−0.658
Gender	←	Character	0.363	0.074	4.881	***	0.348
Traffic convenience	←	Community	1.000				0.846
Public space	←	Community	0.619	0.120	5.166	***	0.527
Leisure space	←	Community	0.399	0.081	4.934	***	0.445
Overall satisfaction	←	Surrounding	1.836	0.361	5.093	***	0.904
Overall satisfaction	←	Socialize	0.627	0.210	2.980	0.003	0.278
Overall satisfaction	←	Community	−0.112	0.070	−1.611	0.107	−0.131
Overall satisfaction	←	Character	0.842	0.245	3.437	***	0.576

***Indicates significance at the $p < 0.001$

In the Daoqian community, potential variables were only "recreation" and "management" (Fig. 7.9). The corresponding observation variables, such as "green landscapes", "public spaces", "recreational spaces", "air quality", "property management", and "pedestrian safety", were shown to have a great impact on satisfaction.

After examining the results of the analysis, we sought to engage in direct dialogue with residents of the Daoqian community to support our understanding of their relative lack of satisfaction in comparison with the residents of Nanhuan. Members of the Daoqian community confirmed the answers given in the questionnaire and declared that they did not think that Nanhuan community provided better living conditions, although they expressed satisfaction with the size of their units. Personal attributes usually affect such assessments, but it is extremely relevant that the residents of Daoqian did not express satisfaction regarding the urban morphology they lived in.

Table 7.9 Modified SEM coefficient results of Daoqian community

			Estimate	S.E	C.R	P	Estimate(S)
Marital status	←	Individual	1.000				0.504
Age	←	Individual	1.498	0.204	7.333	***	0.737
Educational background	←	Individual	−2.755	0.377	−7.299	***	−0.747
Green landscape	←	Recreation	1.000				0.557
Public space	←	Recreation	1.230	0.152	8.078	***	0.742
Leisure space	←	Recreation	0.952	0.122	7.814	***	0.640
Facility diversity	←	Vitality	1.000				0.729
People and cars	←	Vitality	0.735	0.122	6.030	***	0.518
Cultural symbols	←	Vitality	0.745	0.126	5.930	***	0.496
Air quality	←	Management	1.000				0.616
Estate management	←	Management	1.221	0.167	7.322	***	0.646
Pedestrian safety	←	Management	1.143	0.164	6.953	***	0.554
Overall satisfaction	←	Individual	0.087	0.131	0.662	0.508	0.043
Overall satisfaction	←	Recreation	0.494	0.128	3.858	***	0.351
Overall satisfaction	←	Management	0.391	0.144	2.711	0.007	0.242
Overall satisfaction	←	Vitality	−0.117	0.084	−1.398	0.162	−0.110

***Indicates significance at the $p < 0.001$

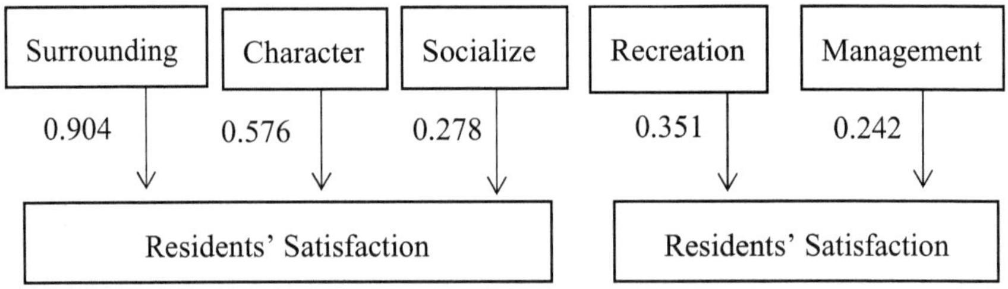

Fig. 7.9 Final influencing factor of Nanhuan (left) and Daoqian (right)

In other words, although their living environment is defined by a completely different urban form, their satisfaction assessments about the living conditions and the size of their units were similar to those given by residents of Nanhuan. The former group was critical of public spaces and community management. Even green spaces were secondary is their assessments, despite the fact that such spaces make up less than 2% of the neighborhood. It can be concluded that the Daoqian community considers itself to be impacted mostly by what is outside of their homes, i.e., recreational spaces and the applied community management model [60, 61].

The above comparison shows significant differences between the residential areas in the two environments. The historic urban area is made up of obsolete, low-rise, high-density blocks inhabited by low-income residents. Furthermore, all aspects concerning living conditions and supporting facilities were found to be far from satisfactory in this middle- class city. Despite the aging population in Suzhou, most of the residents there are long-term residents that enjoy the central location and are used to such a living environment [62, 63].

The Daoqian community is a typical neighborhood in Suzhou and, as such, is subject to historical-cultural protection and development control. Therefore, an acupuncture-style transformation model has to be implemented to improve the conditions and support the creation of recreational and communication spaces. Management has to be improved in order to maintain the area's environmental quality and public security. Finally, although the degree of possible transformation is limited, any action should seek to achieve social sustainable regeneration, that is to say, it should focus on increasing the livability of the local environment, considering which transformations will be accepted and focusing on the revitalization of the area, in order to attract young people and create opportunities and urban vitality [64, 65].

More recent urban areas were built according to modern urban planning theory, in response to the "standardized" demands regarding infrastructure and facilities and green space–high rise settings with large areas being devoted to green spaces and public facilities. The incomes of residents are higher, and their expectations are different from those of the Daoqian community residents.

What is needed in such a setting is to create a differentiated communication open space and to promote transformations focused on residents' perception. Regeneration strategies need to meet the living and psychological needs of different groups, make full use of low land occupation rates, improve environmental protection measures, and to reduce urban noise and air pollution.

7.5 Conclusions

Evaluations of residents' satisfaction have recently become a more prominent part in the Chinese people-oriented regeneration approach to urbanization. This metric allows researchers to measure the efficacy of urban transformations and the regeneration process. This research had the goal of assessing the satisfaction level of the residents of two communities in terms of their quality of life, as it relates to the built environment. Two typical communities with different urban forms were selected as the research objects. Within the framework of community regeneration, the goal of the research was to define a quantitative measurement method by which to obtain the most effective feedback on satisfaction, to understand the main factors which impact resident satisfaction assessments, and to provide suggestions for the design of regeneration initiatives and policymaking [66].

The contribution of this chapter lies in the applied method, i.e., the careful adaptation to case studies of a process that is established in the literature, and in the comparison of the level of satisfaction associated with two different urban morphologies in the same city and the main indicators. This research is original and extremely relevant for the future development in China. On one hand, many projects that were realized after the initial opening-up of the nation are now obsolescent and must be regenerated. Such a process has to be discussed in detail. On the other hand, being socially sustainable implies serious consideration of residents' opinions. In addition to this, the quantity of land used for residential communities, that is to say, the density and compactness of dwellings, must be assessed. Different urbanization models must be taken into consideration and compared. In detail, our comparison of the two communities showed that all residents have the basic need for comfort, safety, and a variety of facilities; however, the specific needs of the two places were found to be different.

Satisfaction in Daoqian community is limited due to the current conditions, and the residents there urgently need new facilities and supporting services. The satisfaction in Nanhuan was found to be more closely related to individual and social network factors, i.e., the improvement of facilities and the diversity of space can significantly improve perceptions by residents of their urban environment. Regeneration initiatives need to create intangible community settings to enhance interactions between residents and the built environment.

Based on these findings, this research makes the following contributions to community regeneration through the prism of social sustainability. First, it establishes a community regeneration conceptual framework that links societies with QoL to achieve sustainable regeneration. Second, according to various factors influencing QoL, it proposes a systematic questionnaire method targeting the individual, built environment, and social interaction levels. The proposed method is of great significance, because there are no unified standards to analyze satisfaction in residential communities. Third, as for the resulting feedback, it was found that the people with different living experiences and individual characteristics and resources express differences in their yearning for space. In drafting policies and regeneration strategies to improve QoL for sustainable development, local satisfaction must be considered.

We were not allowed to obtain the official population data, so we were not able to exactly screen the correlated objects and the number of samples. Therefore, the residents' satisfaction results may have been influenced by one or more additional factors which were not tested. Future research could selectively use and input big data for more accurate analyses.

References

1. State Council of the People's Republic of China (2016) Several opinions on deepening the construction of new urbanization [R] (Document No.: Guo Fa 2016 No. 8). Beijing, State Council. http://www.gov.cn/zhengce/content/2016-02/06/content_5039948.htm. Accessed 30 June 2025
2. State Council of the People's Republic of China (2021) Outline of the 14th five-year plan (2021–2025) for national economic and social development and long-range objectives through the year 2035 [R]. Beijing, Central Party Literature Publishing House. Translated by CCPC Party History and Literature Research Institute. (Official English version)
3. Pellegrini P, Chen J (2020) Hypothesis of densification for a sustainable urbanization in a wealthy Chinese city. IOP Conf Ser Earth Environ Sci 588
4. Zhang X, Chen J, Wang H, Yang D (2025) From Policy Synergy to Equitable Greenspace: Unveiling the Multifaceted Effects of Regional Cooperation upon Urban Greenspace Exposure Inequality in China's Megacity-Regions. Appl Geogr 174:103472. https://doi.org/10.1016/j.apgeog.2024.103472
5. Chen J, Li P, Wang H (2025) Assessment of Influential Factors on Commute and Life Satisfaction in a Historic Campus-Adjacent Environment: Evidence from a Comparison Study of Twin Cities. J Urban Plan Dev 151:4024065. https://doi.org/10.1061/JUPDDM.UPENG-5171
6. Peng Y, Bruyns G, Qu L (2021) Review of Chinese Megablock urbanism: case study of rapid urbanization in the greater bay area. In: Proceedings of the Urbanism in the Mobile Internet Era: 13th Conference of the International Forum of Urbanism, pp 421–430
7. Chen J, Pellegrini P, Ma G (2021) Identifying Resettlement Communities' Urban Regeneration Opportunity Through GIS-Based Spatial Analysis in Suzhou Metropolitan Area. Urban Reg Plan 6:146. https://doi.org/10.11648/j.urp.20210604.15
8. Zhao X, Chen J, Li J, Wang H, Zhang X, Yu F (2025) Unraveling the Renewal Priority of Urban Heritage Communities via Macro-Micro Dimensional Assessment-A Case Study of Nanjing City. China. Sustain. Cities Soc 124:106317
9. Pareja Eastaway M (2012) Social sustainability, vol 6. Elsevier Ltd
10. Karuppannan S, Sivam A (2011) Social Sustainability and Neighbourhood Design: An Investigation of Residents' Satisfaction in Delhi. Local Environ 16:849–870. https://doi.org/10.1080/13549839.2011.607159
11. Diener E, Ryan K (2009) Subjective Well-Being: A General Overview. S Afr J Psychol 39:391–406
12. Mohit MA (2014) Present Trends and Future Directions of Quality-of-Life. Procedia—Soc Behav Sci 153:655–665
13. Ma L, Woods O, Zhu H (2019) Restoration of an ancestral temple in Guangzhou, China: Re-imagining history and traditions through devotion to art and creation. Cult Geogr 26:141–149
14. Chen J, Ren K, Li P, Wang H, Zhou P (2024) Toward effective urban regeneration post-COVID-19: urban vitality assessment to evaluate people preferences and place settings integrating LBSNs and POI. Environ Dev Sustain. https://doi.org/10.1007/s10668-024-05378-8
15. Hopwood B, Mellor M, O'Brien G (2005) Sustainable development: mapping different approaches. Sustain Dev 13:38–52. https://doi.org/10.1002/sd.244
16. Bramley G, Dempdey N, Power S, Brown C, Watkins D (2009) Social sustainability and urban form: evidence from five British cities. Environ Plan A 41:2125–2142. https://doi.org/10.1068/a4184
17. Mohit G, Yadav AK, Bisht NS, Giresh M (2008) Valuation of recreational benefits from valley of flowers national park. Indian For 134:26–35
18. Zhan D, Meng B, Zhang W (2014) A study on residents' perception of housing satisfaction and behavioral intention in Beijing. Geogr Stud 2:336–348
19. Lucht F, Frey B, Salmons J (2011) A tale of three cities: urban-rural asymmetries in language shift? J Ger Linguist 23:347–374
20. Carey EC, Covinsky KE, Lui L, Eng C, Sands LP, Walter LC (2010) Prediction of mortality in community-living frail elderly people with long-term care needs. J Am Geriatr Soc 56:68–75

21. Liu Z, Wang X, Ma J (2020) The impact mechanism of Beijing community public space on neighborhood communication in the transitional period: a comparative analysis of local residents and immigrants. Geogr Sci 40:69–78
22. Stevenson B, Wolfers J (2013) Subjective well-being and income: is there any evidence of satiation? Am Econ Rev 103:598–604
23. The state council 1aaguiding opinions of the general office of the state council on comprehensively promoting the renovation of old urban communities
24. Chan E, Lee GKL (2008) Critical factors for improving social sustainability of urban renewal projects. Soc Indic Res 85:243–256. https://doi.org/10.1007/s11205-007-9089-3
25. Littig B, Griessler E (2005) Social sustainability: a catchword between political pragmatism and social theory. Int J Sustain Dev 8:65–79
26. Liu J, Tan X, Cheng Q (2017) Practice and thinking of participatory community planning under the background of transformation—a case study of y community, Qinghe street, Beijing. Shanghai Urban Plan 6
27. Canter D (1977) The psychology of place. Psychol Place
28. Muchadenyika D, Waiswa J (2018) Policy, politics and leadership in slum upgrading: a comparative analysis of Harare and Kampala. Cities 82:58–67
29. Jeon JY, You J, Chang HY (2007) Sound radiation and sound quality characteristics of refrigerator noise in real living environments. Appl Acoust 68:1118–1134
30. Ma B, Zhou T, Lei S, Wen Y, Htun TT (2019) Effects of urban green spaces on residents' well-being. Environ Dev Sustain 21:2793–2809
31. Zhang Q, Hiu Kwan Yung E, Hon-Wan Chan E (2021) Meshing sustainability with satisfaction: an investigation of residents' perceptions in three different neighbourhoods in Chengdu, China. Land 10:1280
32. Du H (2002) Factors influencing residents' satisfaction with residential environment quality—a case study of several residential areas in the pearl river delta. Urban Plan Forum 7
33. Lee K.-Y (2022) Relationship between physical environment satisfaction, neighborhood satisfaction, and quality of life in Gyeonggi, Korea. Earth Its Atmos 10:663
34. Cao J, Zhang J (2016) Built environment, mobility, and quality of life. Travel Behav Soc 5:1–4
35. Diener E, Scollon CN, Lucas RE (2009) The evolving concept of subjective well-being: the multifaceted nature of happiness. In: Diener E (ed) Assessing well-being: The collected works of Ed Diener. Springer Science + Business Media, pp 67–100
36. Zhao M, Liu N, Chen J, Wang D, Li P, Yang D, Zhou P (2024) Navigating post-COVID-19 social–spatial inequity: unravelling the nexus between community conditions, social perception, and spatial differentiation. Land 13
37. Hu J, Chen J, Li P, Yan J, Wang H (2024) Systematic review of socially sustainable and community regeneration: research traits, focal points, and future trajectories. Buildings 14:881. https://doi.org/10.3390/buildings14040881
38. Li G (2015) Survey on residents' satisfaction with living in typical areas of Shenzhen. Urban Issues 72–77
39. Chen J, Gan W, Liu N, Li P, Wang H, Zhao X, Yang D (2024) Community quality evaluation for socially sustainable regeneration: a study using multi-sourced geospatial data and AI-based image semantic segmentation. ISPRS Int J Geo-Inf 13(5):167. https://doi.org/10.3390/ijgi13050167
40. Chen J, Zhao X, Wang H, Yan J, Yang D, Xie K (2024) Portraying heritage corridor dynamics and cultivating conservation strategies based on environment spatial model: an integration of multi-source data and image semantic segmentation. Herit Sci 12:419. https://doi.org/10.1186/s40494-024-01497-7
41. Chen J, Li P, Wang H (2024) MSPA-based collaborative evaluation and planning governance of territory and natural environment in metropolitan agglomerations—a case study of egiddz in yangtze river delta. J Resour Ecol 15:1263–1273. https://doi.org/10.5814/j.issn.1674-764x.2024.05.014
42. Song H, Chen J, Li P (2025) Decoding the cultural heritage tourism landscape and visitor crowding behavior from the multidimensional embodied perspective: insights from Chinese classical gardens. Tour Manag 110:105180. https://doi.org/10.1016/j.tourman.2025.105180

43. Wang J (2016) A study of living satisfaction in the urban village relocation areas: the case of Zhengzhou urban villages. Archit J 14:86–89
44. Yuan Y, Ding K, Cao X, Wu X (2018) A review of neighborhood satisfaction. Urban Dev Stud 25:111–117
45. Riazi M, Emami A (2018) Residential satisfaction in affordable housing: a mixed method study. Cities 82:1–9. https://doi.org/10.1016/j.cities.2018.04.013
46. Tan Y, He J, Han H, Zhang W (2019) Evaluating residents' satisfaction with market-oriented urban village transformation: a case study of Yangji village in Guangzhou, China. Cities 95:102394. https://doi.org/10.1016/j.cities.2019.102394
47. Huang X, He D, Tang S, Li X (2020) Compensation, housing situation and residents' satisfaction with the outcome of forced relocation: evidence from urban China. Cities 96:102436. https://doi.org/10.1016/j.cities.2019.102436
48. Yi L, Jianping L (2024) The influence factors of residential satisfaction of urban floating population in China. Chongqing Soc Sci 61–68
49. Niu J, Zhao M (2018) Analysis on factors influencing of yurts residence satisfaction based on structural equation model. Sci Dev 14:750–756
50. Wang P, Qin X, Li Y (2021) Satisfaction evaluation of rural human settlements in northwest China: method and application. Land 10:813
51. Sang Z, Xia S (2013) Community awareness: interpersonal relationship, social embeddedness and community satisfaction—a survey of urban residents' community identity. Nanjing Soc Sci 7
52. Yang N, Chen J, Ban L, Li P, Wang H (2024) Pre-planning and post-evaluation approaches to sustainable vernacular architectural practice: a research-by-design study to building renovation in Shangri-La's Shanpian House, China. Sustain 16:9568
53. Chen J, Wang H, Yang Z, Li P, Ma G, Zhao X (2023) Comparative spatial vitality evaluation of traditional settlements based on SUF: taking Anren ancient town's urban design as an example. Sustain 15:8178
54. Chen J, Pellegrini P, Yang Z, Wang H (2023) Strategies for sustainable urban renewal: community-scale GIS-based analysis for densification decision making. Sustain 15:7901
55. Chen J, Li P, Lei Y, Zhang Y, Lai C, Chen B, Liu J, Schnabel MA (2025) Leveraging augmented reality for historic streetscape regeneration decisionmaking: a big and small data approach with social media and stakeholder participation integration. Cities. https://doi.org/10.1016/j.cities.2025.106214
56. Anderson JC, Gerbing DW (1988) Structuralequation modeling in practice: a review and recommended two-step approach. Psychol Bull 103:411
57. Ma G, Paola P, Wu H, Han H, Wang D, Chen J (2023) Impactof land-use mixing on the vitality of urban parks: evidence from big data analysis in Suzhou, Yangtze river delta region. China. J Urban Plan Dev 149:4023045. https://doi.org/10.1061/JUPDDM.UPENG-4334
58. Chen J, Tian W, Xu K, Pellegrini P (2022) Testingsmall-scale vitality measurement based on 5D model assessment with multi-source data: a resettlement community case in Suzhou. ISPRS Int J Geo-Inf 11:626. https://doi.org/10.3390/ijgi11120626
59. Kearns A, Mason P (2013) Defining and measuring displacement: is relocation from restructured neighbourhoods always unwelcome and disruptive? Hous Stud 28:177–204
60. Zhang X, Lu Y, Xu Y, Zhou C, Zou Y (2024) Governing regional inequality through regional cooperation? A case study of the Guangdong-Hong Kong-Macau greater bay area. Appl Geogr 162:103135. https://doi.org/10.1016/j.apgeog.2023.103135
61. Zhang X, Zhou Z, Song Y (2024) Contested geopolitics of distribution in the city-region-building: promoting enclave industrial parks in the pearl river delta, China. Trans Inst Br Geogr e12697
62. Zhang X, Sun Y, Luan X (2023) Regulation flexibility and legitimacy building in governing intercity railways: the polymorphous role of the Chinese provincial government. J Urban Aff 1–16
63. Li Y, Zhang X, Xia C (2023) Towards a greening city: how does regional cooperation promote urban green space in the Guangdong-Hong Kong-Macau greater bay area? Urban For & Urban Green 86:128033

64. Zhang X, Shen J, Sun Y, Zhou C, Yang Y (2023) Formation of city regions from bottom-up initiatives: investigating coalitional developmentalism in the pearl river delta. Ann Am Assoc Geogr 113:700–716
65. Zhang X, Chen Y, Xu Y, Yue W, Tang C (2023) Does regional cooperation constrain urban sprawl? Evidence from the Guangdong-Hong Kong-Macao greater bay area. Landsc Urban Plan 235:104742
66. Pellegrini, Paola; Chen, Jinliu The Regeneration Path Not Taken Further: An Experiment in Urban Densification and State Entrepreneurialism in a Resettlement Neighborhood in Suzhou, China. Urban Stud. 2025, 00420980251365188, doi: https://doi.org/10.1177/004209802513 65188.

Chapter 8
Densification Testing via Vitality Measurement Based on 5D Model Assessment with Multi-Source Data in Nanhuan

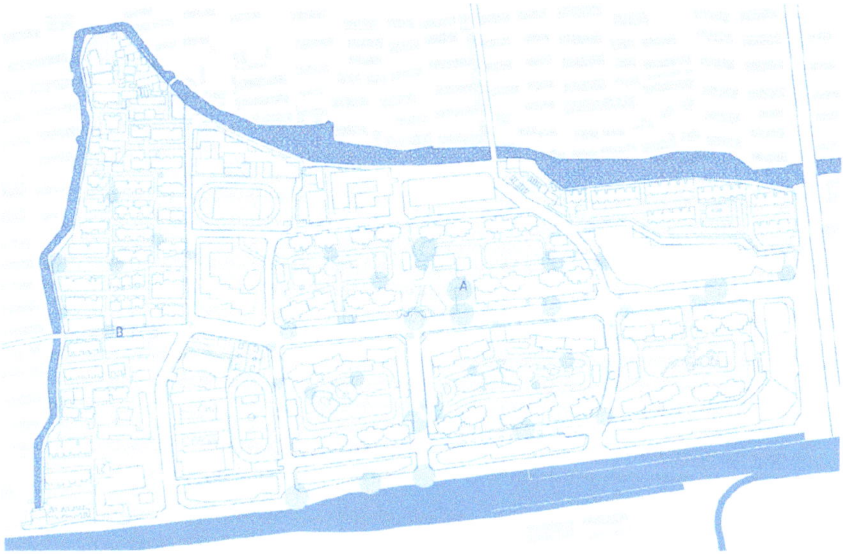

8.1 Introduction

8.1.1 Community Vitality Enhancement: Regeneration Movement Orientation in China

In recent years, critical reflection on the rapid urbanization of previous decades in China has been gradually increasing. Responding to the low spatial quality and vitality under unprecedented urbanization, the 14th five-year national plan (2021–2025) first mentioned urban regeneration movement for reactivating the existing

J. Chen and P. Pellegrini, *Densification as a Design Tool for Sustainable Urban Regeneration*, Urban Sustainability, https://doi.org/10.1007/978-981-95-1653-7_8

urban environment as an essential goal to optimize urban fabric and promote high-quality spaces with the guidance of people-oriented and harmonized living principles [1]. This new concept is highly associated with restoring the profitability and vitality of existing communities [2]. The urban renewal intentions it contains include a range of plans for demolition and redevelopment, partial renovation and full-scale transformation, and these latest guidelines align China with the direction of sustainable development [3]. Academics and practitioners are emphasizing how to measure and create vibrant urban spaces effectively [4]. The renovation of old and dilapidated housing estates, which are the basic units of social space, is an important step towards high-quality urbanization [5].

Urban vitality, as has been highlighted in various theories and literature, plays a key role in meeting people's demands for high-quality living and promoting urbanization connotative development [6]. Jacobs defines a vibrant urban environment as an assembly of urban design elements, including appropriate development density, land use and functional mix, aged buildings, small blocks, accessible streets, and a lack of border vacuums. Lynch describes a "supportive urban environment" as a settlement supported by vital functions and capacities of human beings. Supported by the provision of vibrant urban spaces, the well-being of people and the development of long-term, sustainable urban regeneration will facilitate a wide range of human activities, encourage social contact and interaction, and foster a feeling of safety and community among residents [7]. Therefore, precisely identifying the status quo of community vitality is much needed in formulating reactive and sustainable urban policies and designing interventions in the new context of connotative growth.

Information and communication technology development brings about tremendous available open-source geographic data, which has pushed the study of quantifying urban vitality into an unprecedented era [8, 9]. With the penetration of quantitative assessment methods into urban design disciplines, spatial data mining and GIS (Geographical Information System) analysis techniques are being successfully applied to explore various planning elements related to vibrant urban places. Taking advantage of big data, Fan used hourly Baidu heat map (BHM) data as a reference to depict urban vitality along the Yangtze River in Nanjing [10]. Wang employed big data gathered from street view imaging (SVI), points of interest (POI), and social media comments to explore and discuss street space quality, providing a novel alternative for evaluating public realms based on the interplay between street vitality, service facilities, and the physical environment [7]. Guo evaluated the street's vitality by adopting optimized K-means clustering to identify street dynamic vitality categories and assessing the classification result based on vitality intensity and stability [11].

Urban vitality represents the quality of a place in terms of the interaction level between human activities and spatial entities [12]. Scholars define spatial vitality as inclusiveness, aggregation density, and urban park spatial usage intensity [13]. Li described spatial vitality as the degree of satisfaction related to humans' basic movements while living, producing, and exploring their surroundings [14]. Vibrant cities are an assembly of good forms, developed functions, and sufficient activities [15]. Therefore, various qualitative methods have been proposed to evaluate different planning elements, including urban morphology, population density, facility density, plot size, the degree of mix use, and landscape quality [16]. Determinants such as attractiveness, diversity, and accessibility have also been investigated, measured, and proved positively correlated with urban vibrancy [17, 18]. These previous studies have guided urban planning practice on a city or regional scale. Nevertheless, there is limited evidence to determine whether these pre-existing measurement metrics for spatial vitality are transferrable to neighborhood regeneration, partly due to the dilemma of land use, complex regeneration policy, and property issues. Multi-disciplinary knowledge is required to assist in making adaptive urban development strategies and design interventions [19, 20].

8.1.2 New Urban Quantitative Analysis via Multi-Source Data and Wearable Equipment: Measuring Community Vitality

A quantitative analytic framework is an excellent tool for assessing spatial or spatiotemporal urban vitality to conduct the multi-criteria urban analysis of the status quo. Inspired by theoretical research, academics have methodically and comprehensively examined various spatially explicit distribution features and the implicit driving mechanisms of the built environment. In previous studies, scholars have classified the characteristics of urban space into economic, cultural, and social dimensions and generated multi-factor frameworks to synthesize urban dynamics as a weighed overlay result of economic, social, and environmental indicators [21]. This paper summarizes these frameworks into the following two categories.

The first category of indicators focuses on allocating and developing infrastructure, business, and service resources. Kunze and Hecht extracted small catering businesses' geographic distributions and impact to estimate area-based urban vitality [22]. A JPL (Junctions, POI, Location-based Services) appraisal matrix has been proposed to assess urban spaces from the aspects of traffic, function, and activity density [23].

The second category is more associated with urban morphological features. In recent years, numerous studies have shown that human activity intensity, the physical built environment, and human–environment interaction profoundly impact urban vitality [24]. Cervero and Kockelman suggested the potential impacts of urban design on urban vitality, including intersection density and mixed-use land [25]. Ye used regression models to explore how density and typology have influenced the spatial vitality of Shenzhen [26]. Li further classified spatial vitality into social, economic, and cultural dimensions, and interpreted them with street-related data to determine how the external environment has shaped the spatial vitality of Chengdu [27]. The '5D' theory (Density, Diversity, Design, Destination accessibility, Distance to Transit) has more extensive coverage as well as a more refined measurement that incorporates all the relevant concepts of urban vitality [28, 29]. Apart from typical morphological features, it focuses on walking behavior that can represent community residents' convenience, quality, and health status in localized scenarios [30]. Vale & Pereira stressed the efficacy of BE (built environment) factors under the '5D' dimensions with medium-sized Portuguese cities taken as study areas [31]. Ogra & Ndebele (2013) regarded '5D' components as the primary components that constitute citywide transit-oriented development (TOD) [32].

Although the preceding studies provide practical tools for quantifying the degree and elements of spatial vitality from holistic dimensions, there are still problems with providing robust evidence for community regeneration. Firstly, there is a lack of universal measurement to synthesize spatial activity at different scales or across different geographic boundaries, which has added more difficulty in precisely illustrating the differences in the spatial distribution of human–environment interactions on the micro-scale. Secondly, recent studies have only laid a technical foundation for measuring urban vitality with the '5D' model in large-scale urban areas. The ways to tackle scaling-down issues, such as the repeated pattern of BE within a single phase of property development, the lack of detailed resolution (<30 m) satellite images, higher impurities, and poorer positioning accuracy in the construction of local human activities, have also become significant research constraints. Thirdly, a limited number of studies have focused on micro-scale measurement of spatial elements that shape high-quality communities or provide implemental strategies to guide community regeneration. Due to these difficulties, field observations, interviews, or other participatory methods are mainly adopted for primary data acquisition in prevalent Chinese community regeneration studies [33, 34]. These difficulties have constrained the current community regeneration within its narrow scope of building façade renewal, infrastructure enhancement, or other similar "beautification" tactics, without the interference of data insight. However, these "placebo" strategies have insufficient contributions to resolving deep-rooted structural problems or improving the genuine living standards of local citizens. To fill the gap between urban regeneration practices and urban vitality assessment, it is essential and promising to conduct systematic and reactive local vitality evaluation at the community level.

In recent years, diverse methods for tracking human activity in big data have received widespread attention. With the development of sensor technologies, a substantial real-time record of the relationships between human activities and the built environment has emerged in recent research [35]. In other words, the time, the sequence of daily activities, and the geographic location with specific features of land use and population density can be used to infer spatial or spatiotemporal urban vitality. Some studies have utilized conventional environmental sensors to record people's activities and various ground space characteristics. However, the expense and expertise required for deployment and maintenance of the monitoring system make it more challenging to achieve the density level required for refined community monitoring.

As opposed to traditional sensor techniques, street images have demonstrated superior efficiency and geographic reach for gathering micro-scale street features and pedestrian volume associated with the '5D' urban analysis assessment framework and community vitality reflection [36]. Our work uses a mobility-based technique using wearable camera picture collection equipment to gain refined community vitality based on image identification and analysis software, while addressing flaws such as poor reliability owing to the area coverage and update frequency. Timely-taken pictures can provide high-quality data sources of precise scale, high validity, and fast and stable location updates for spatial vitality identification and verification due to the small size, lightweight, and portability of wearable cameras [37, 38]. This provides the basis for perceiving more complex and three-dimensional community environment information and comprehensively evaluating the community vitality from the space–time perspective. However, the long collection period, high cost, and rigid schedules for field observation in the construction of the human activity process have made this approach hard to promote citywide, which could be seen as a significant research challenge [39, 40].

In summary, the urgent agenda for evidence-based urban vitality improvement and Chinese community regeneration, data collection and mining technologies development, as well as the prevalence of quantitative analytical methods, provide a new paradigm for current urban scientific research [41, 42]. However, the existing analytical framework has not only failed to provide universal measurements to synthesize spatial vitality, but has also neglected to explore the application of research models at the community or local scale. This research combines multi-source heterogeneous data to provide a fundamental evaluation framework of urban vitality performance and its internal driving force based on the '5D' theory (i.e., design, diversity, density, distance to transit, and destination accessibility). It seeks to provide a comprehensive and systematic approach for measuring the connotation of local-scale spatial vitality [43, 44]. According to our best knowledge of previous research, this framework presents the following three innovations:

1. It is a novel paradigm for urban vitality study based on the deconstruction of connotation and explicit explanation of driving mechanisms. The study seeks to broaden the contemporary '5D' theory assessment framework to guide small-scale community vitality assessment. The measurement criteria selected for each dimension are elaboratively tailored to local-scale planning components and purposefully moderated to better manifest the spatial vitality to solve the "scaling-down" issues.
2. Tracking human activities and inferring vitality differences with timely site photography is more accurate and efficient than traditional collection methods.
3. The proposed framework fully uses multi-source data and offers a multi-dimensional methodology for examining urban vitality driving indicators. This research investigates the underlying driving mechanism of urban vitality by combining the morphological and perceptual perspectives for shaping urban vitality efficiency.

Our framework enriches the theoretical framework, quantitative methodologies, and assessment variables of urban vitality. It reveals the method for developing urban vitality from the viewpoint of community regeneration. It offers multiple insights into enhancing urban vibrancy and supporting the coordinated and sustainable development of a declining Chinese high-density neighborhood. As the global urbanization forces continue to accelerate, the gravitational center of human dynamics continues to shift to East Asia, particularly China [45]. Therefore, studies of urbanization and its countermeasures require additional Chinese experience. This research examines a resettlement community in Suzhou, China, to verify the framework's practical applicability. Based on our research on the driving mechanism of spatial vitality, we summaries the techniques for boosting urban vitality in terms of collaborative policymaking and urban design [3, 46].

The remainder of this article is structured as follows. The following part describes the evaluation framework's conceptual framework, data requirements and methodology, the case study area, and experimental configurations. In the subsequent sections, conclusions are discussed, followed by an elaboration of case study results.

8.2 Analysis Framework

Community vitality is crucial to constructing vibrant urban environments and an important element of sustainable urban regeneration that improves the quality of life (QoL). However, rapid urbanization in China has caused significant urban problems, such as the proliferation of inactive urban and community spaces [47, 48]. Incorporating community vitality into urban regeneration agendas demands reactive urban regeneration strategies and appropriate design interventions to address new problems [49, 50]. Therefore, it is essential to develop a precise and comprehensive community vitality assessment framework to maximize regeneration potentials, assist in policy formulation, and urban design, and improve QoL to the most significant extent. The two primary components of vitality evaluation in the present study area are qualitative studies based on questionnaires and interviews, and quantitative studies based on urban data and network analysis. Although researchers have begun applying multi-dimensional digital data to the urban analytics area, the constraints of conventional approaches continue to pose substantial challenges to accurately measuring community and local vitality. With the iteration of new technologies, urban vitality assessment studies have entered a prosperous period marked by an abundance of open-source geospatial data. According to the above thoughts, this research proposes a general analysis and evaluation methodology for measuring community vitality based on the '5D' model assessment with multi-source data.

As shown in Fig. 8.1, our study examines two distinct categories of data. The first type consists of built environment (BE) data, including point of interest (POI) data, land use, building, and street network, which are the data source for evaluating the '5D' planning elements. The second category is composed of image data, which manifests the real-time distribution pattern of local vitality. Three core steps constitute our research: the urban vitality measurement, evaluation, and verification.

Part 1 shows the urban vitality driving factors designed for this framework. We formulate twelve evaluation indicators to measure urban spatial vitality separately from the '5D' dimensions (Density, Diversity, Design, Destination accessibility, and Distance to transit). The framework chooses evidence-based quantitative assessment

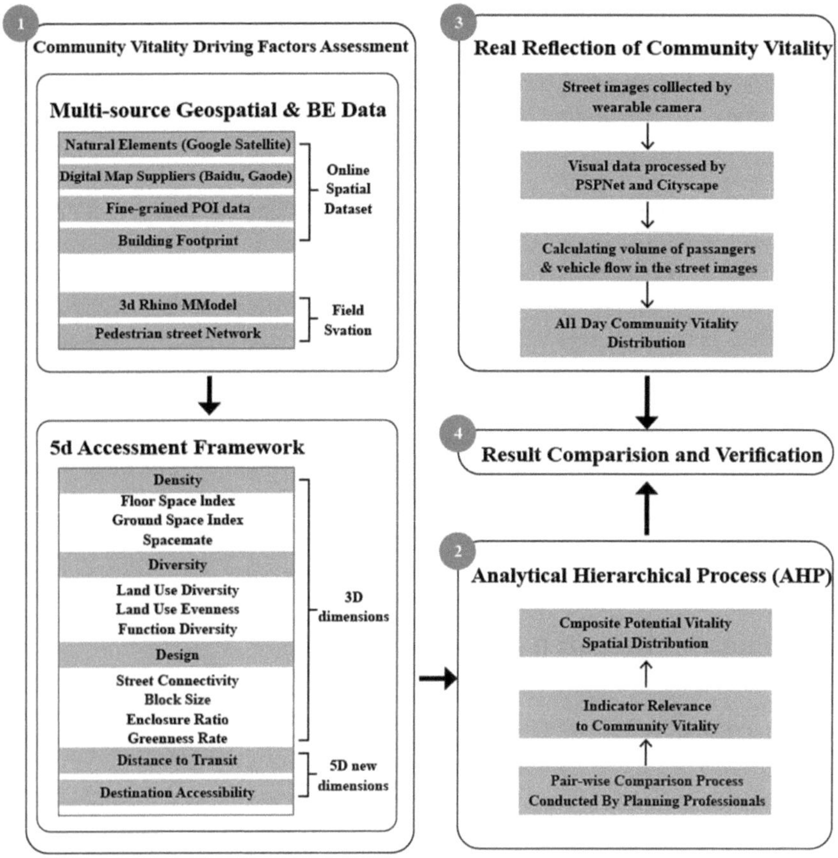

Fig. 8.1 The conceptual framework of community vitality evaluation

to reflect the connotations of each dimension. High-density neighborhoods tend to accommodate more activities within the same commuting radius. They are more concerned with the morphological features of individual blocks, which the floor space index (FSI) measures, Ground Space Index (GSI), and Spacemate [51]. Land use and functional (POI) diversity represents the distribution of location-based activity and land use practices, and is an essential catalyst for community vibrancy [12]. As the design dimension reflects pedestrians' walking experiences and the aesthetic aspects of urban space, street connectivity, block size, greenness index, and enclosure ratio are the four selected variables for the measurement of mobility, attractiveness, and dynamics of urban spaces [52, 53]. Closer distance to transit increases residents' willingness to travel [54], and high destination accessibility boost walking activity, urban mobility, and sustainability [31].

As seen in Part 2, we employ the Analytical Hierarchical Process (AHP) to accomplish two goals. First, we use a hierarchical matrix to compare the '5D' influencing factors in terms of their contribution to community vitality. Second, we compute the composite impacts of '5D' variables on the distribution of community vitality (by block) by assigning weights to each depending on their relevance to community regeneration. The community scale requirements have been incorporated into the weight allocation process to improve the efficiency and consistency of the weighing process.

Part 3 describes the actual measurement and verification technique for vitality. Our methodology proposes a new perspective based on improved real-time data-gathering technologies. Especially when environmental sensing, image crowd-gathering, and quantitative recognition techniques are prevalent, the flexibility and precision of relevant devices become increasingly crucial for effective data collection and analysis. We use the PSPNet algorithm and Cityscape to process visual data collected from wearable cameras, and use the volume of passengers and vehicle flow in the street images to reflect community vitality.

In Part 4, by comparing it with the results calculated from the '5D' theory assessment, our research's contribution to the quantitative community vitality assessment methodology and real-life application is verified.

8.3 Data and Research Methods

8.3.1 Data Source

According to our framework, multi-source data is the technical foundation of this paper. Image data covering accessible areas collected from a 15-day site tour, and BE data acquired from multi-source geospatial databases, are fused by individual urban blocks. They are used to identify the distribution of urban vitality and the driving factor for modelling and assessment. The composition, usage, source, and contents of these data are displayed in Table 8.1.

8.3.2 Quantification Calculation of '5D' Community Vitality Impact Factors Precise GIS Spatial Vitality Analysis Based on '5D' Theory Assessment Model

According to the conception of the '5D' theory, several key indicators are selected from the previous application of the '5D' theory, and moderated to suit the community scale. They are used to evaluate the two-dimensional ground value of urban ground BE features and human–land interaction. The meaning and calculation of the indicators are illustrated in Table 8.2.

In the first three formulas, F is the floor height, and L denotes a designated block's average number of floor layers. Ax and Bx represent the areas of the individual block

Table 8.1 The input datasets used in this framework

Item	Usage	Contents	Source
Crowd-gathering Street image data	Community vitality real reflection and verification	The number of pedestrians and vehicles, collected by fixed-interval photography and processed by image recognition software	Xiaotu S2 Sports Camera during the 20-time site visits
Geospatial big data	The '5D' driving indicators measurement and assessment	A detailed base map covering the whole community, including the information of: Pedestrian-accessible Road networks Building footprints with height Urban plot with an area Fine-grained POI data, modified according to real contextual conditions	Various navigation map suppliers, including Open Street Map, Google Satellite Map, Baidu Map, Gaode Map, etc
Statistical unit for data fusion	Basic supporting data	The urban block is set as a basic unit for '5D' driving factors measurement and assessment. The boundaries are determined based on satellite images and government planning documents	3D modeling using CAD and Rhinoceros
Basic analysis tools	Community vitality real reflection and verification	A suitable size regular grid for reflecting the real-life spatial distribution of community vitality	Manually create fishnet using ArcGIS

and the gross regions of all building footprints. The FSI shows the average density of 3d built-up spaces or population density. GSI reflects the average density of 2d ground floor spaces, described as 'built potential' [26]. According to some scholars think that higher vitality derives from higher FSI/GSI, and GSI tends to exert a greater impact on urban vitality [55, 56].

The mixing index of the Shannon entropy function represented in Eq. (4) is widely used to measure spatial diversity and functional richness based on the variety of site types and the proportion of each type. LD refers to land use diversity. P_k is the proportion of the site area of the type in k to the block area, while n is the number of land use types in the block. A higher land use density for each block represents a better mix of facilities. In Eq. (5), LE refers to the land use evenness, LD_k refers to the land use diversity of the block, while LD_{max} is the maximum value of the land use diversity among all the blocks. The closer the value is to 1, the more evenly distributed the block in the site. In Eq. (6), FD denotes functional density, N_k indicates the total number of POIs in the block, and A_k denotes the area of the block. FD reflects the potential of a region to provide different types of urban services, with the higher value representing the higher potential of block functions. Jiang have proven that LD, LE,

Table 8.2 Quantification calculation method for community vitality impact factors designed in our framework

Decision Objectives in Scheme Level	Indicators in the First Intermediate Layer	Equation	Number	Data Source	Description
Density	Ground space index (GSI)	GSI = Bx/Ax	(1)	Building footprints with height and plot area data	Referred to the Spacematrix by Berghauser Pont & Haupt (2005)
	Floor space index (FSI)	Fx = F × L FSI = Fx/Ax	(2)		
	Spacemate	/	(3)		
Diversity	Land use diversity (LD)	$LD = -\sum_{k=1}^{n} P_k ln(P_k)$	(4)	POI and plot area data	Referred the Shannon entropy method by Jiang (2021) and Dong, Zhang (2016), and Kunze and Hecht (2015)
	Land use evenness (LE)	$LE = \frac{LD_k}{LD_{max}}$	(5)		
	Function density (FD)	$FD = \frac{N_k}{A_k}$	(6)		
Design	Link connectivity (LC)	LC = Street Connectivity Index in SDNAs	(7)	Pedestrian accessible road networks	LC evaluates: street density the ratio of minor streets to major streets
	Greenness index (GI)	/		Street photos taken at fixed points	The average values of the proportion of greenness area/ enclosed areas from all crowd-gathering photos are the inputs
	Enclosure rate (ER)	/			
	Block size (BS)	/		Plot area data	"Quantifying Jacobs" (1961) theory

(continued)

Table 8.2 (continued)

Decision Objectives in Scheme Level	Indicators in the First Intermediate Layer	Equation	Number	Data Source	Description
Position	Destination accessibility (DA)	DA = Local Integration Index in SDNAs	(8)	Pedestrian-accessible route networks	The calculation uses Space Syntax with GIS SDNAs plug-in
	Distance to transit (DT)	/		Navigation map data	The shortest distance between each block reflects it to the bus stop

and FD positively impact urban vitality. Nevertheless, the three indices contribute to urban vitality to varying degrees [57].

We use the "link connectivity index" to access street connectivity with the Space Syntax plug-in in ArcGIS. In Eq. (7), higher link connectivity (LC) represents higher community vitality. The sampling databases for the greenness index (GI) and enclosure rate (ER) are built upon the graphic information taken regularly at 54 data points of 20-time site visits for two consecutive weeks. The photos were divided into different combinations of built-up areas, greenness areas, road areas, cars, and pedestrians by proportion. GI refers to the proportion of green features in a photo, which is positively correlated with urban vitality. Oliveira and Medeiros (2016) proposed using the ratio of "building height to street width, namely enclosure ratio, to delineate how built-up elements can influence pedestrian walking experience and suggest a comprehensive method to measure the effect of average building setbacks, space between facades and street width simultaneously [58]. Therefore, we use the road proportion divided by the building proportion to calculate ER, which is negatively associated with urban vitality.

Destination accessibility (DA) and distance to transit are combined (DT) as one Decision Objective in Scheme Level: Position. We used the Local Integration Index in Spatial Design Network Analysis (SDNAs) to study the DA of the Nanhuan New Village. To avoid the boundary effect caused by the selecting area, the study selected rivers and highways as a boundary, which have apparent blocking effects on pedestrian movement. A higher DA results in higher local selectivity of the space unit within a certain travel radius, and the higher the space vitality can be obtained. We measure the shortest walking distance from the exit of each residential area to each bus stop to obtain the DT for each block. Areas with less DT are considered to have higher vitality.

8.3.3 Weighed Evaluation and Composition of Community Vitality Based on Analytic Hierarchy Process (AHP)

The Analytical Hierarchy Process (AHP) is a decision-making tool that helps to solve complex problems through problem decomposition, criteria selection, pairwise/relative comparison, and synthesis of the relative importance of rankings [59, 60]. The analytical hierarchy process has been widely used in assessing and interpreting urban regeneration elements [61, 62]. We use it to evaluate the '5D' influencing factors in terms of their contribution to community vitality and calculate the total vitality index of individual blocks. Experts assess the elements of '5D' decision objectives and weight attribution results are ranked in scheme level and in the first intermediate layer.

Table 8.3 shows the 12 selected indicators that make up the '5D' decision objectives. To make each factor comparable, the 12 types of data and their impacts (+/−) are remapped to the interval of values between 0 and 1, and weighed based on the results from AHP. "Density, Diversity, and Design" have been considered more responsible for community vitality and thus obtained higher weights, i.e., 0.4236, 0.2270, and 0.2270, respectively. Compared with the three significant factors, position-related variables ('Distance to transit' and 'Destination Accessibility') have lower weights (0.1223) due to the constrained scale of a community. The weights of the indicators vary from 0.0236 to 0.2286. The following formula gives the Potential Vitality Index (PVI) for the AHP approach. D is the number of possible factors from the '5D' dimensions, and X is the weight of each factor.

$$\text{Potential Vitality Index (PVI)} = \sum_{i=1}^{9} D_i \times X_i$$

Table 8.3 AHP assigned weight for urban vitality calculation

Decision objectives in scheme level	Weight in scheme level	Indicators in the first intermediate layer for decision objectives	Weight in the intermediate layer	Positive or negative correlation
Density	0.4236	Ground space index (GSI)	0.2286	+
		Floor space index (FSI)	0.1258	+
		Spacemate	0.0692	+
Diversity	0.2270	Land use diversity (LD)	0.0568	+
		Land use evenness (LE)	0.1135	+
		Function diversity (DF)	0.0568	+
Design	0.2270	Link connectivity (LC)	0.0929	+
		Block size (BS)	0.0236	–
		Greenness index (GI)	0.0731	+
		Enclosure rate (ER)	0.0375	–
Position	0.1223	Destination accessibility (DA)	0.0815	+
		Distance to transit (DT)	0.0408	–

8.3.4 *Reflection and Verification of Community Vitality Based on the Wearable Camera*

Our mobility-based approach to estimating local vitality is based on visual data collection through the wearable camera and data procession via PSPNet (Pyramid Scene Parsing Network) [63]. The spatially explicit characteristics of urban vitality are extracted and compared with the '5D' assessment results by the following three components:

1. Two volunteers, equipped with wearable cameras, were invited to carry out 20-time experiments for two consecutive weeks (once during the daytime, once in the evening). The sampling database is built on street images taken regularly at 54 designated points within the site boundary.
2. The continuous timing photos are put into the PSPNet to calculate the ratio of vehicles and pedestrians in each image. The time–space differences of this ratio reflect the distribution of local spatial vitality.
3. These outcomes can be compared with the vitality distribution acquired from the '5D' assessment framework to validate the effectiveness of our proposed methodology, especially a static approach using geospatial data and mobility-based methods employing street view photos.

In this study, the real-time distribution of local spatial vitality is reflected by the time–space differences of the streetscape features that the pedestrians can directly perceive, specifically by extracting the proportion of people and vehicles using PSPNet [64]. The pyramid pooling module and the pyramid scene parsing network are efficient machine-learning approaches for scene parsing and semantic segmentation. This method has obtained pixel-level prediction performance on several datasets, such as Cityscapes [65]. As shown in Fig. 8.2, five types of ground items (built-up area, green space, road area, car and pedestrian) were extracted from each street view picture for each ground item using a pre-trained Cityscapes model. Then, we use the composite ratio of pixels of pedestrians and automobiles to the total number of pixels in the picture to represent the vitality of a community, since pedestrian and vehicular traffic tend to be associated with a geographical concentration of human activities. Lastly, we measure the average composite ratios of all sampling points in 20-time walking experiments to obtain the time–space distribution of community vitality in all possible environmental conditions.

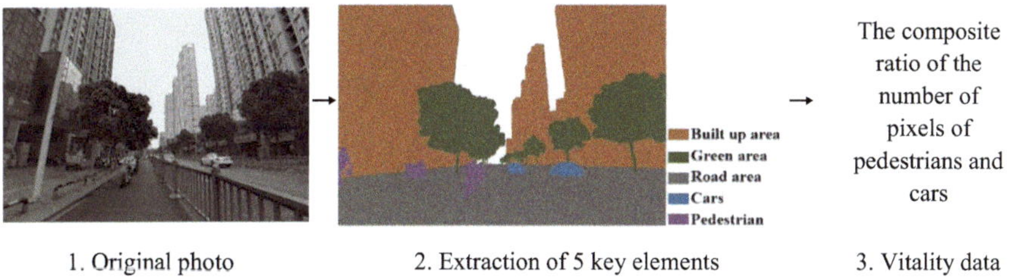

Fig. 8.2 The image procession approach based on PSPNet and Cityscapes

8.4 Case Study and Experimental Design

8.4.1 Case Study Area and Experimental Data

In response to the double pressure of land resource scarcity and economic development, Chinese urban regeneration has embraced the high-rise, high-density residential development pattern [66]. A '3H' community refers to a densely populated community with a high Floor Area Ratio (FAR, indicating the intensity of plot use), high building height, and high occupancy rate. In current years, '3H' development has led to excessive residential land use, which is a source of traffic congestion, environmental quality deterioration, a lack of green open space, and severe air pollution [67].

We selected Nanhuan New Village as our study area, which is a typical '3H' community (Table 8.4). This area is located in the southern part of Gusu District, Suzhou City, Jiangsu Province, China. Suzhou functions as the hinterland of the Taihu Lake Economic Development Zone and has a strategic position in the Yangtze River Delta City Cluster. Moreover, Gusu District is a historic core and a highly urbanised area, where the embodiment of urban vitality and the formation mechanism of urban regeneration can be fully reflected.

The geographical location, natural and administrative boundaries, and the built-up elements of the study area are shown in Fig. 8.3. The west and north of the community

Table 8.4 Basic information of '3h' Nanhuan New Village

Indicators	Nanhuan New Village
Total land area	21.54 ha
Gross Floor Area (GFA)	555,375.76 m^2
Residential area and percentage	391,425 m^2
Floor Area Ratio (FAR)	2.58
Building density	26.8%
Greenness rate	25.6%
Residential units	4852
Parking lots	2526

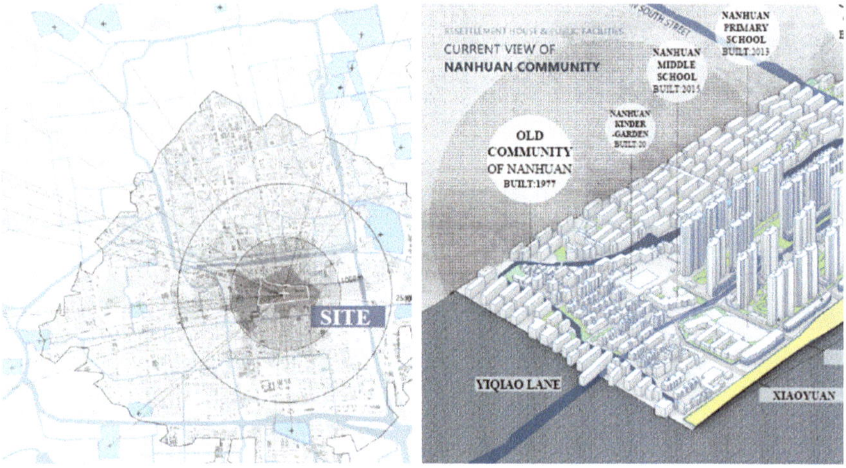

(a) Nanhuan geographic location in Suzhou (b)Nanhuan community 3D model rendering

Fig. 8.3 The geographic location of the study area, area boundaries, and contextual information

are surrounded by water. The eastern boundary is adjacent to Nanyuan South Road, and Nanhuan Highway demarcates the Southern boundary. The site is composed of the 1970s-built old neighborhood in the west, a kindergarten, a primary school, a middle school in the middle, and newly built (post-2010) high-rise, high-density residential blocks that dominate the rest of the area. The southeastern outer ring of residential blocks is surrounded by retail streets and has a large shopping mall. According to the urban vitality evaluation framework demonstrated in Sect. 8.4.1, street image data and multi-source geospatial big data were obtained as research data. Their specific descriptions are detailed in Table .8.5

For user activity type inference needs, the original categories of POI are merged into 24 categories. They are rental services, lottery, photography, laundry, stationery, pedicure, real-estate agency, office, repairing, food retails, fashion, gym, travel agency, advertisement agency, food wholesales, tobacco and wine, education, restaurant, bank, convenience store, flower shop, appliance shop, baby caring and vacant outlet.

The picture in Fig. 8.3 shows the location of Suzhou city, and Fig. 8.3b shows the distribution of resettlement communities in Suzhou city. The 3D model shows the building footprints, road networks, and neighborhood blocks in our study area at a finer scale.

8.4.2 Experimental Settings

To facilitate the efficiency of visual data collection and programming process, and to verify the distribution of community vitality more effectively, we used street images

Table 8.5 A detailed description of the data source used in our empirical research

Item	Source	Quantity	Time
Street image data	Field observations with Xiaotu S2 Sports Camera	3029 photos	1–15 August 2021
Building footprints with height Urban blocks and building plots profile	Google Satellite Map, the world's largest spatial information supply (* Modified after field observations) Accessed from: https://maps.google.com/, accessed on 1 August 2021	97 polygons	Accessed on August 2021
Pedestrian-accessible road networks	Baidu Map and Gaode Map, China's most extensively used map service platform (* Modified after field observations) Accessed from: https://map.baidu.com/, accessed on 1 August 2021	1479 lines	
Fine-grained POI data	Open Street Map, the world's leading free wiki world map (Modified after field observations) Accessed from: https://www.openstreetmap.org/, accessed on 1 August 2021	68 points	

collected from accessible areas as an alternative proxy of overall spatial vitality. We divided the study area into 17 urban blocks as basic spatial analysis units and design two routes along with 54 navigation points (Fig. 8.4). The two routes encompass all the plots and paths accessible to pedestrians as a valid and comprehensive picture of space activities [68]. The 54 points were set in a continuous sequence with explicit starting and ending positions to help us better match all the street images with their geographic coordinates. Dashed lines indicate the areas with comparable spatial vitality distribution to the designated route 1.

Then, we asked two volunteers to walk along the routes with a wearable camera to capture street images for 15 consecutive days. They were asked to maintain a pedestrian average walking speed (4–5 km/h) and the time threshold of a single stay $\Delta t = 2$ h. Since the route spreads throughout Nanhuan New Village, the data collected during the stay can be used to explain the vitality distribution of the entire community. In this study, we chose the Xiaotu S2 Sports Camera for its outstanding portability and stability that fully fulfil our experiment's required flexibility and preciseness. We set the scenario for the Xiaotu S2 Sports Camera to take a photo every 10 s, indicating that it has a relatively high temporal resolution and quick reflection for tracking the participants' surrounding environmental features, such as green features, cars, and people. The records of their travels were divided into routes and points.

Our experiment was conducted from 1 to 15 August 2021. Specifically, each participant was instructed to wear a camera to capture the front view during the daytime over two consecutive weeks, from 8:00–10:00 and 18:00–20:00. Before the walking session, the volunteers would ensure that the front lens was sturdy, facing forward, unobstructed, and fully charged. During the experiment, participants were

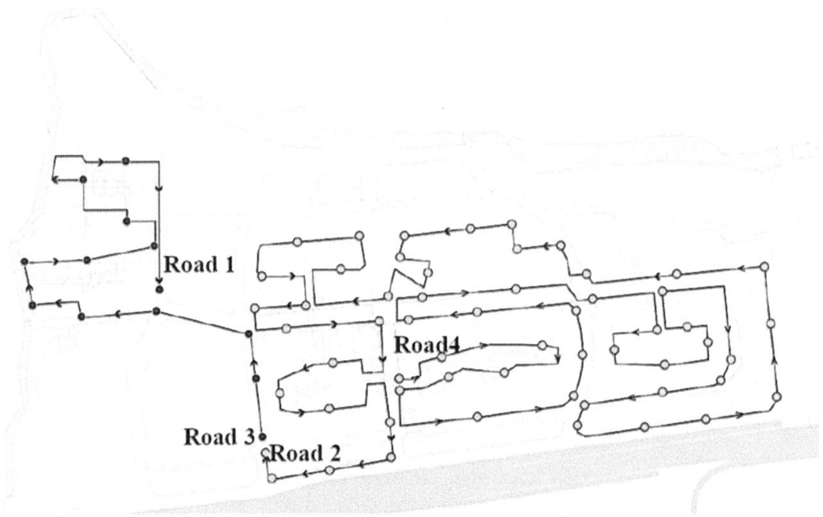

Fig. 8.4 The image data collection routes and the 54 navigation points

responsible for ensuring that their wearable cameras were functioning properly unless they were engaged in an activity or event unsuitable for exposure. Meanwhile, the volunteers' privacy concerns were taken into consideration. They were allowed to remove the camera at any moment for privacy concerns and could delete the images containing private information. After each day of walking observation, volunteers were asked to remove the wearable cameras and export the images to our mobile hard drives.

8.5 Results

8.5.1 Quantitative Results of Vitality

The quantitative results of the 12 indicators affecting vitality in the 5D model are shown in Fig. 8.5. For comparison purposes, the evaluation results were classified into ten levels using the natural breakpoint classification method.

1. The graphs (density1-density3) show that both GSI and FSI are generally more significant in the residential areas than in the public and commercial land areas, with the FSI and Space mate values in the central high-rise significantly higher than in the older mid-rise residential areas.
2. In terms of diversity (diversity1-diversity3), the spatial distribution of LD and LE is similar, with high indicators for high-rise residential areas with commercial areas in the center and low indicators for older mid-rise single residential areas.

The southern neighborhood near the South Circular Road has a higher DF value due to the variety and density of POI points. This reflects the complete diversity in the central and southern parts of the neighborhood.

3. Regarding design (design1-design3), the LC has a more prominent spatial differentiation related to the entrances and exits of each plot. The residential areas to the northwest and the residential areas to the southeast are areas of high LC con-centration, while the other regions are lower. The distribution of BS values is similar to that of density, again with the residential areas having greater values than the public and commercial land areas. The GI gradually increases from west to east, indicating that the high-rise residential and commercial areas to the east are better landscaped. The ER is highest in the east and smallest in the central neighborhoods with squares. This reflects the lower degree of enclosure of the public areas.
4. Regarding position (position1 &2), the school areas have the highest DA values, followed by the commercial areas, reflecting their higher accessibility. The distribution of DT shows that the residential areas are all at shorter distances from public transport stations, reflecting the accessibility of the neighborhood.

We employed the AHP to compare the '5D' influencing factors in terms of their weight on community vitality and compute the composite impacts of '5D' variables on the potential distribution of community vitality. The first column (A-Q) represents the 17 blocks in Nanhuan New Village based on the street cut-out. The second to 13th columns (GSI-DT) represents the quantified results of the 12 indicators in the 5D model for each block, calculated from Table 8.2. The last column (PVI) indicates the potential vitality of each block based on the 12 indicators quantified using the AHP method. As described in Table 8.3, the weights of density, diversity, design, and position are 0.4236, 0.2270, 0.2270, and 0.1223, respectively, so the final vitality was obtained by multiplying each quantitative indicator by its corresponding weight and adding them together ($PV = \sum_{i=1}^{9} D_i \times X_i$). The data in Fig. 8.5 are derived from Table 8.6 and visualised on the map using GIS.

According to the results of the PVI in Table 8.6, the distribution of the Vitality Index across the blocks in Nanhuan New Village is shown in Fig. 8.6. The community vibrancy in Nanhuan New Village is highest in the central high-rise mixed commercial and residential area, decreasing in order of radioactivity towards the periphery. The high-rise strip blocks (I, J, K, O) with double-story street retail carry the highest community vitality values. These blocks are centrally located on Nanhuan Street, with highly connected streets and easy access to local amenities. Not only do the super-high residential buildings accommodate a large population, but the user-friendly retail also brings a diversity of site mix and functions. This is closely followed by the neighborhood-type commercial use areas (P, Q) to the east, which has higher GSIs, denser POI densities, and smaller block sizes, despite their seemingly distant location from the residential areas to the west. However, not all neighborhood-type commercial building plots have the same vitality level. F, L, and M along the South Ring Elevated Wa, on the other hand, have low densities, lack diversity, and low levels of vitality due to the massive closure of many shops (Fig. 8.7). The areas with lower

Fig. 8.5 Visual estimation results of the 12 indicators affecting vitality in the 5D model

vitality levels are concentrated in the westside school districts (C, D, E) and the old Nanhuan community (A, B, N). The former has less social or economic activity and a higher level of enclosure due to land use. For the latter, the old Nanhuan community occupies a larger land area and is designed as a single-function, medium to high-rise residential area. It is also somewhat removed from the main centres of economic and social activity in the centre and east and is, therefore, a low-vitality area.

8.5.2 Urban Vitality Based on Sensory and Visual Data Traced by Wearable Camera

Armed with portable sensors and wearable cameras, the research team conducted 20 field conditions collection sessions in Nanhuan New Village in both the morning and the afternoon. One of the direct contributions to urban vitality was the observation of pedestrian range and flow. The research team aggregated the information collected

Table 8.6 AHP evaluation results for Potential Vitality Index (PVI)

Block	GSI	FSI	Spacemate	LD	LE	DF	LC	BS	GI	ER	DA	DT	PVI
A	0.035	0.099	0.046	0.005	0.007	0.013	0.032	0.027	0.085	0.023	0.051	0.032	0.45960
B	0.023	0.064	0.0461	0.008	0.048	0.097	0.037	0.025	0.079	0.010	0.048	0.032	0.52355
C	0.024	0.067	0.0461	0	0	0	0.028	0.025	0.088	0.004	0.081	0	0.36621
D	0.024	0.079	0.0461	0	0	0	0.026	0.026	0.092	0.003	0.081	0	0.38027
E	0.031	0.087	0.0461	0	0	0	0.035	0.023	0.077	0.010	0.072	0	0.38546
F	0.054	0.184	0.0692	0.018	0	0	0.066	0.025	0.074	0.003	0.044	0	0.54262
G	0.032	0.155	0.0692	0.002	0.029	0.059	0.043	0.023	0.083	0.006	0.056	0	0.56120
H	0.042	0.178	0.0692	0.002	0	0	0.051	0.022	0.082	0.005	0.067	0	0.52155
I	0.124	0.087	0.0231	0.014	0.056	0.113	0.047	0.025	0.080	0.019	0.062	0.021	0.67764
J	0.113	0.105	0.0231	0.027	0.050	0.099	0.037	0.029	0.081	0.013	0.058	0.021	0.66175
K	0.104	0.087	0.0231	0.024	0.047	0.095	0.040	0.029	0.084	0.013	0.065	0.021	0.63720
L	0.039	0.164	0.0692	0.056	0	0	0.073	0.028	0.074	0.003	0.063	0	0.57206
M	0.042	0.178	0.0692	0.052	0	0	0.060	0.027	0.077	0.004	0.049	0	0.56288
N	0.025	0.055	0.0461	0.001	0.022	0.045	0.053	0.031	0.029	0.006	0.077	0.040	0.43520
O	0.125	0.076	0.0231	0.017	0.043	0.086	0.048	0.034	0.086	0.008	0.061	0.024	0.63610
P	0.068	0.228	0.0692	0.015	0	0	0.057	0.037	0.085	0.023	0.057	0	0.64284
Q	0.038	0.161	0.0692	0.035	0.028	0.055	0.050	0.030	0.078	0.019	0.053	0	0.62036

GSI Ground space index, *FSI* Floor space index, *LD* Land use diversity, *LE* Land use evenness, *DF* Function diversity, *LC* Link connectivity, *BS* Block size, *GI* Greenness index, *ER* Enclosure rate, *DA* Destination accessibility, *DT* Distance to transit, *PVI* Potential Vitality Index

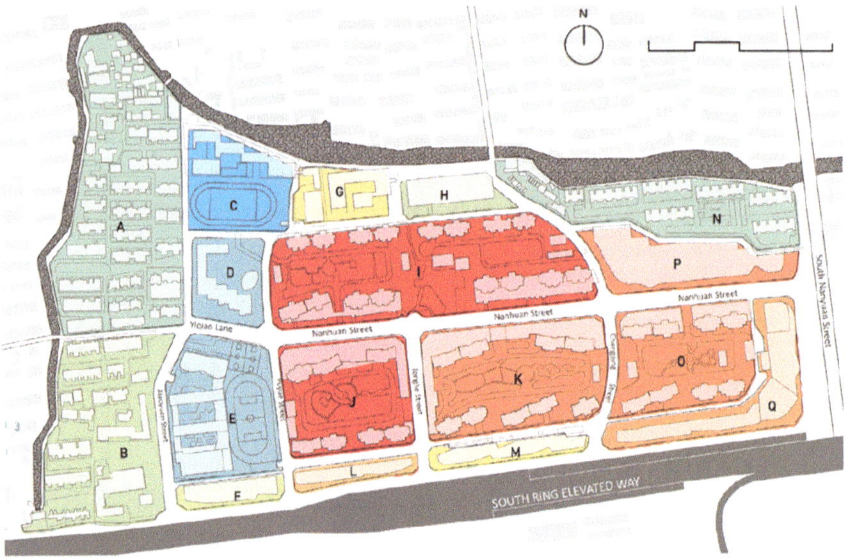

Fig. 8.6 Distribution map of potential vitality in Nanhuan New Village

Fig. 8.7 Closed shops on the roadside

and used GIS to map the distribution of pedestrian flows at different times of the day, as shown in Fig. 8.8. The proximity principle was applied to compare the vitality values of the area covered by the route with those of the surrounding communities to determine and visualize overall distribution, including some inaccessible areas.

Figure 8.8 shows that the range and volume of pedestrian flows observed during the day are higher than in the afternoon, suggesting that crowd activity is more active in the morning than in the afternoon in Nanhuan New Village. Areas of significant difference are mainly located along the roads of the school and in the commercial areas along the streets in the high-rise residential areas to the southeast of the community. This relates to residents' morning habits of picking up children and grocery shopping. Throughout the day, the junction with the square in the center of the community and the shopping area to the east has the highest flow of people. This reflects the high level of vitality of the central high-rise residential area and the eastern commercial area as the most active areas of social and economic activity.

Central Square is a local activity hub with a high crowding level. This functional open space allows people to organize activities such as shopping, dancing,

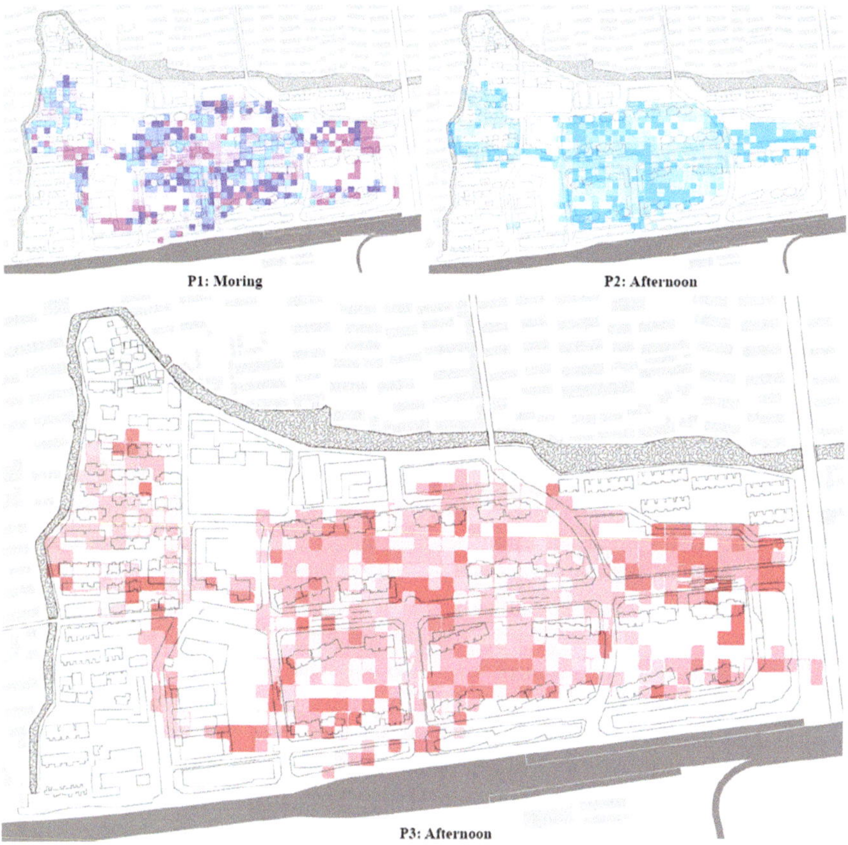

Fig. 8.8 Distribution map of pedestrian flows

fitness and chatting. In addition to the squares, pedestrians are also present in each neighborhood's green spaces and entrances. People prefer the former for recreational activities, such as resting, and have to enter and leave the community through the latter. In summary, the degree to which people congregate is significantly influenced by the type of activities the place can serve locally. More specifically, an area's land use and diversity represent this capacity, and we argue that diversity contributes significantly to the measurement of urban vitality when using the 5D model.

8.6 Discussion

8.6.1 *How Can the '5D' Model Guide Small-And-Medium Scale Community Vitality Assessment?*

Combing the vitality distribution results from the '5D' model assessment and previous literature, we can summaries the following conclusions. By comparing all the driving factors, 'Density' has the most noticeable positive impact (especially the GSI and FSI indicators, which are 0.2286 and 0.1258, respectively). In contrast, block size has no significant driving impact on urban vitality in Nanhuan New Village (as shown in Table 8.7, BS has been given to only 0.0236 of weight and therefore shows a minimal impact). The most vibrant blocks are those of larger block sizes, and the smallest-sized school districts obtain a relatively low PVI. Some blocks (A & I; Q & E) have similar size but have disparate vitality values. Additionally, although the weight given to 'Diversity' and 'Design' is similar, the former demonstrates a greater association with urban vitality than the latter. These results, in general, coincided with the '3d' assessment model, as higher levels of population density and built potentials, functional diversity and mix-use, and high-quality urban design are prerequisites for stimulating superior vitality performance [4, 17, 33].

The difference between this article and that of Li et al. (2022) is that their research did not consider the weight for each '5D' indicator, and they listed both position factors ('Destination accessibility' and 'Distance to transit') as the two most important factors driving urban vitality [27]. However, at the community level, the original '3d' dimensions ('Density', 'Diversity', 'Design') remain key elements for forming vibrant spatial quality and functionality, while the position factors have relatively lower effects. This indicates that despite the great importance of mobility factors in transport planning (e.g., TOD), community regeneration should focus more on improving the internal environment and the morphological features of living space. This echoes the research conclusions on urban design factors of spatial vitality [24]. The most significant difference between this article and previous research outcomes is the enclosure rate. According to some scholars think that a low enclosure rate can bring about a better walking experience, whilst in the study area, mixed-use and high-density urban areas are located next to the streets enclosed by high-rise blocks, which might have offset the negative impact of enclosure rate on community vitality[58].

Previously, the '5D' model has only been used to measure vitality performance across large-scale urban areas. At the same time, this paper fills the gap between community regeneration practices and urban vitality assessment by proposing a refined '5D' model. Based on a deep reflection of previous studies, the measurement criteria, analysis, and weighting process are selected and moderated to conduct systematic and reactive local vitality evaluation at the neighborhood level.

Table 8.7 Comparison between theoretical and realistic scenarios of '5D' indicators on urban vitality

Dimension	Indicator	Theoretical correlation	Real-life correlation	AHP Weight	AHP Weight by dimension
Density	Ground space index (GSI)	+	+	0.2286	0.4236
	Floor space index (FSI)	+	+	0.1258	
	Spacemate	+	+	0.0692	
Diversity	Land use diversity (LD)	+	+	0.0568	0.2270
	Land use evenness (LE)	+	+	0.1135	
	Function density (DF)	+	+	0.0568	
Design	Link connectivity index (LC)	+	+	0.0929	0.2270
	Block size (BS)	–	Irrelevant	0.0236	
	Green index (GI)	+	+	0.0731	
	Enclosure rate (ER)	–	+	0.0375	
Position	Destination accessibility (DA)	+	+	0.0815	0.1223
	Distance to transit (DT)	+	+	0.0408	

8.6.2 How Can Real-Time Street Image Support Accurate and Efficient Data Collection?

Supported by new technologies and data, this study is a valuable attempt to combine human-scale elements with urban-scale analysis and effectively extend the analysis techniques for community vitality. Compared to traditional vitality assessments that rely on GIS overlay analysis, this study provides a more comprehensive and systematic approach based on a human perspective. Due to a lack of appropriate tools and data, previous vibrancy studies and practices have often been conducted from a top-down perspective with a flat analysis, making it challenging to incorporate relevant elements of the human scale. Nanhuan New Village, as a case study in this research, has the characteristics of a small-scale and complex habitat. This study is based on the human and environment-oriented perspective to address this characteristic. The wearable camera was used to measure the real-time spatial quality of the street, pedestrian, and vehicular information, compensating for the shortcomings of traditional quantitative analysis in terms of small scale and accuracy. Content capture of street photos taken by wearable cameras enabled greenery visibility and pedestrian numbers to be obtained more accurately. Meanwhile, the combination of street-based data analysis and spatial design network analysis (sDNA) software allowed for accurately measuring street network accessibility.

In summary, wearable devices can be used in conjunction with the collection and use of urban streetscapes to address the shortcomings of quantitative urban research that ignores the personality and characteristics of the city itself. This multi-source analysis of urban data can be used to study issues at the city scale and community level with a focus on the human dimension. This is particularly suitable for research applications at the micro and medium scales. With the use of photographic location information as the starting point, and software visualization technology as the support, the research process is spatially recorded. The image elements are managed orderly, offering the possibility of in-depth and accurate analysis using the 5D concept to measure the quality of the built environment.

Using wearable cameras to collect street images helps tackle issues such as homogeneous BE patterns within a single phase of property development, the lack of high-resolution satellite images, higher impurities, and poorer positioning accuracy in the construction of local human activities. Therefore, our research proposes an effective way to overcome significant research constraints in urban data analysis, which is more accurate and efficient than traditional data collection and processing methods.

8.6.3 How to Verify the Assessment Results from Morphological and People- and Environment-Oriented Perspectives?

The '5D' assessment results and the actual distribution of urban vitality have apparent spatial heterogeneity. By comparing Figs. 8.6 and 8.8, we can obtain two explicit conclusions. First, the diversity level calculated from both methods is highly overlapped in the retail streets (Tonghe Street, central Nanhuan Street) and the eastern periphery of the Huilin Centre shopping mall. Central Health Square and Tonghe Street intersection shows the most significant crowd gathering and the highest PVI. The street shops provide diverse places for lingering, consumption, and community gathering, and the large shopping mall also stimulates high urban vitality.

Unlike Tonghe Street, the intersection of two neighborhood streets (Yiqiao Lane and Xiaoyuan Street) shows disparate vitality levels in the two methods. The reason could be that the place is located at the only entrance of Nanhuan New Village. Another difference is the informal entrance of the community, located between blocks F and L (Fig. 8.6), on the western segment of the South Ring Express Way of the study area. These entrances directly bring about large flows of pedestrians and cars. Their volume concerns employment and industrial links between the community and the city. Thus, such inconsistency is not caused by evaluation units or problematic processing data approaches and does not impair the reliability of the proposed assessment model.

Overall, the modelling and assessment process results precisely reflect real-life community vitality. Therefore, this comparative study combines the shaping effect of morphological features and people and environment-oriented perspectives to offer a new methodology for community-level vitality assessment. It successfully proves the validity of the '5D' model and multi-source big data's validity in effectively examining urban vitality and its driving indicators.

8.7 Conclusion

The development of information collection technologies has changed the way scholars' study urban science. Wearable devices, such as cameras and sensors, for example, provide practical tools for the human-centered perspective of urban identity mapping and new exploration in the context of digital urban renewal. Meanwhile, spatial network analysis tools such as GIS and sDNA are gradually developing and maturing, and scholars in urban research are gradually applying multi-source urban data.

As envisaged by Ewing and Cervero, the quantification of urban vitality can be reflected in terms of density, diversity, design, destination accessibility, and transportation distance [69, 70]. This study effectively integrates classical 5D theory with urban data and weighting algorithms in depth to propose a spatial vitality measure for high-density resettlement communities with a general practical application. Through data from field works and calculations, this study shows how the level of vitality changes in the small scale high-density residential community selected as a case study. In detail Tonghe Street, central Nanhuan Street, and the eastern periphery of Huilin Centre have very high levels of vitality, which is also verified by the crowd gathering and activities measured in Nanhuan New Village. This higher vitality is due to the morphological features of living space.

Based on these findings, our study proposes some suggestions for measures to improve the vitality of such small-scale, high-density communities. The methodology and findings can be extended to other regions or cities with similar characteristics and can be used to develop urban vitality shaping strategies [71].

Acknowledgments The author gratefully acknowledges Wenkang Tian (Harvard University) and Kexin Xu (UCL) for their contributions to data acquisition, analysis, visualization, as well as their insightful input during writing and revision of earlier versions of the research.

References

1. Fang CL (2019) Basic rules and key paths for high-quality development of the new urbanization in China. Geogr Res 38:13–22
2. Ruming K (2018) Urban regeneration in Australia. Routledge, Oxon
3. Chen J, Pellegrini P, Wang H (2022) Comparative residents' satisfaction evaluation for socially sustainable regeneration—the case of two high-density communities in Suzhou. Land 11:1483. https://doi.org/10.3390/land11091483
4. Zhang X, Chen J, Wang H, Yang D (2025) From policy synergy to equitable greenspace: unveiling the multifaceted effects of regional cooperation upon urban greenspace exposure inequality in China's megacity-regions. Appl Geogr 174:103472. https://doi.org/10.1016/j.apgeog.2024.103472
5. Chen J, Pellegrini P, Xu Y, Ma G, Wang H, an Y, Shi Y, Feng X (2022) Evaluating residents' satisfaction before and after regeneration. the case of a high-density resettlement neighbourhood in Suzhou, China. Cogent Soc Sci 8:0–26. https://doi.org/10.1080/23311886.2022.2144137
6. Lu S, Huang Y, Shi C, Yang X (2019) Exploring the associations between urban form and neighborhood vibrancy: a case study of Chengdu, China. ISPRS Int J Geo-Inf 8:165
7. Wang T, Li Y, Li H, Chen S, Li H, Zhang Y (2022) Research on thevitality evaluation of parks and squares in medium-sized chinese cities from the perspective of urban functional areas. Int J Environ Res Public Health 19:152–238
8. Kang C, Fan D, Jiao H (2021) Validating activity, time, and space diversity as essential components of urban vitality. Environ Plan B Urban Anal City Sci 48:1180–1197. https://doi.org/10.1177/2399808320919771
9. Chen J, Li P, Wang H (2025) Assessment of influential factors on commute and life satisfaction in a historic campus-adjacent environment: evidence from a comparison study of twin cities. J Urban Plan Dev 151:4024065. https://doi.org/10.1061/JUPDDM.UPENG-5171
10. Fan Z, Duan J, Luo M, Zhan H, Liu M, Peng W (2021) How did built environment affect urban vitality in urban waterfronts? a case study in Nanjing reach of Yangtze river. ISPRS Int J Geo-Inf 10:611
11. Guo X, Chen H, Yang X (2021) An evaluation of street dynamic vitality and its influential factors based on multi-source big data. ISPRS Int J Geo-Inf 10:143
12. Jiang Y (2021) A study on urban vitality measurement and its spatial characteristics—a case study of the main urban area of Wuhan. Urban Manag Sci Technol 22:4
13. Hui EC man, Chen T, Lang W, Ou Y (2021) Urban community regeneration and community vitality revitalization through participatory planning in China. Cities 110:103072
14. Li X, Wen X (2021) Analysis on urban vitality based on spatial syntax. Urban Rural Plan 18:12–17. https://doi.org/10.19892/j.cnki.csjz.2021.15.02
15. Den Hartog H (2010) Shanghai new towns: searching for community and identity in a sprawling metropolis. 010 Publishers, ISBN 906450735X
16. Liu S, Zhang L, Long Y, Long Y, Xu M (2020) A new urban vitality analysis and evaluation framework based on human activity modeling using multi-source big data. ISPRS Int J Geo-Inf 9. https://doi.org/10.3390/ijgi9110617
17. Chen J, Ren K, Li P, Wang H, Zhou P (2024) Toward effective urban regeneration post-COVID-19: urban vitality assessment to evaluate people preferences and place settings integrating LBSNs and POI. Environ Dev Sustain. https://doi.org/10.1007/s10668-024-05378-8
18. Zhao M, Liu N, Chen J, Wang D, Li P, Yang D, Zhou P (2024) Navigating post-COVID-19 social–spatial inequity: unravelling the nexus between community conditions, social perception, and spatial differentiation. Land 13
19. Hu J, Chen J, Li P, Yan J, Wang H (2024) Systematicreview of socially sustainable and community regeneration: research traits, focal points, and future trajectories. Build 14:881. https://doi.org/10.3390/buildings14040881
20. Chen J, Gan W, Liu N, Li P, Wang H, Zhao X, Yang D (2024) Community quality evaluation for socially sustainable regeneration: a study using multi-sourced geospatial data and AI-based

image semantic segmentation. ISPRS Int J Geo-Inf 13(5):167. https://doi.org/10.3390/ijgi13050167
21. Pugalis L (2009) The Culture and economics of urban public space design: public and professional perceptions. Urban Des Int 14:215–230. https://doi.org/10.1057/udi.2009.23
22. Kunze C, Hecht R (2015) Semantic enrichment of building data with volunteered geographic information to improve mappings of dwelling units and population. Comput Environ Urban Syst 53:4–18. https://doi.org/10.1016/j.compenvurbsys.2015.04.002
23. Jin X, Long Y, Sun W, Lu Y, Yang X, Tang J (2017) Evaluating cities' vitality and identifying ghost cities in china with emerging geographical data. Cities 63:98–109. https://doi.org/10.1016/j.cities.2017.01.002
24. Ye Y, Huang R, Zhang L (2021) Human-oriented urban design with support of multi-source data and deep learning: a case study on urban greenway planning of Suzhou Creek, Shanghai. Landsc Arch 28:39–45. https://doi.org/10.14085/j.fjyl.2021.01.0039.07
25. Cervero R, Kockelman K (1997) Travel demand and the 3DS: density, diversity, and design. Transp Res Part D Transp Environ 2:199–219. https://doi.org/10.1016/S1361-9209(97)00009-6
26. Ye Y, Zhuang Y (2017) A hypothesis of urban morphogenesis and urban vitality in newly built-up areas: analyses Basedon street accessibility, building density and functional mixture. Urban Plan Int 32:43–49
27. Li Q, Cui C, Liu F, Wu Q, Run Y, Han Z (2022) Multidimensional urban vitality on streets: spatial patterns and influence factor identification using multisource urban data. ISPRS Int J Geo-Inf 11. https://doi.org/10.3390/ijgi11010002
28. Ewing R, Cervero R (2001) Travel and the built environment: a synthesis. Transp Res Rec 1780:87–114
29. Ewing R, Cervero R (2010) Travel and the Built environment: a meta-analysis. J Am Plan Assoc 76:265–294
30. Lu Y, Xiao Y, Ye Y (2017) Urban density, diversity and design: is more always better for walking? a study from Hong Kong. Prev Med (Balt) 103:S99–S103. https://doi.org/10.1016/j.ypmed.2016.08.042
31. Vale DS, Pereira M (2016) Influence on pedestrian commuting behavior of the built environment surrounding destinations: a structural equations modeling approach. Int J Sustain Transp 10:730–741. https://doi.org/10.1080/15568318.2016.1144836
32. Ogra A, Ndebele R (2013) The role of 6DS: density, diversity, design, destination, distance, and demand management in transit oriented development (TOD). Neo-Int Conf Habitable Environ 539–546
33. Chen J, Zhao X, Wang H, Yan J, Yang D, Xie K (2024) Portraying heritage corridor dynamics and cultivating conservation strategies based on environment spatial model: an integration of multi-source data and image semantic segmentation. Herit Sci 12:419. https://doi.org/10.1186/s40494-024-01497-7
34. Chen J, Li P, Wang H (2024) MSPA-based collaborative evaluation and planning governance of territory and natural environment in metropolitan agglomerations—a case study of EGIDDZ in Yangtze river delta. J Resour Ecol 15:1263–1273. https://doi.org/10.5814/j.issn.1674-764x.2024.05.014
35. Goos G, Hartmanis J, Van J, Board LE, Hutchison D, Kanade T, Kittler J, Kleinberg JM, Kobsa A, Mattern F et al (2010) Human behavior understanding. First International Workshop, Turkey
36. Chen L, Lu Y, Ye Y, Xiao Y, Yang L (2022) Examining the association between the built environment and pedestrian volume using street view images. Cities 127:103734
37. Pan L, Gan W, Chen J, Xian X (2024) Ecological planning for mitigating urban heat based on thermal network simulation: a case study of Guangzhou-Foshan metropolitan area. J Chin Urban For 22:10–19. https://doi.org/10.12169/zgcsly.2024.06.26.0004
38. Yang N, Chen J, Ban L, Li P, Wang H (2024) Pre-planning and post-evaluation approaches to sustainable vernacular architectural practice: a research-by-design study to building renovation in Shangri-La's Shanpian House, China. Sustain 16:9568

39. Chen J, Wang H, Yang Z, Li P, Ma G, Zhao X (2023) Comparative spatial vitality evaluation of traditional settlements based on SUF: taking Anren ancient town's urban design as an example. Sustain 15:8178
40. Chen J, Pellegrini P, Yang Z, Wang H (2023) Strategies for sustainable urban renewal: community-scale GIS-based analysis for densification decision making. Sustain 15:7901
41. Zhang X, Lu Y, Xu Y, Zhou C, Zou Y (2024) Governing regional inequality through regional cooperation? a case study of the Guangdong-Hong Kong-Macau greater bay area. Appl Geogr 162:103135. https://doi.org/10.1016/j.apgeog.2023.103135
42. Zhang X, Zhou Z, Song Y (2024) Contested geopolitics of distribution in the city-region-building: promoting enclave industrial parks in the pearl river delta, China. Trans Inst Br Geogr e12697
43. Pan L, Gan W, Chen J, Ren K (2023) An integrated model for constructing urban ecological networks and identifying the ecological protection priority: a case study of Wujiang District, Suzhou. Sustain 15:4487
44. Ma G, Paola P, Wu H, Han H, Wang D, Chen J (2023) Impact of land-use mixing on the vitality of urban parks: evidence from big data analysis in Suzhou, Yangtze river delta region, China. J Urban Plan Dev 149:4023045. https://doi.org/10.1061/JUPDDM.UPENG-4334
45. Berawi MA, Ibrahim BE, Gunawan G, Miraj P (2019) Developing a conceptual design of transit-oriented development to improve urban land use planning. J Des Built Environ 19:40–48 https://doi.org/10.22452/jdbe.vol19no1.4
46. Chen J, Li P, Lei Y, Zhang Y, Lai C, Chen B, Liu J, Schnabel MA (2025) Leveraging augmented reality for historic streetscape regeneration decisionmaking: a big and small data approach with social media and stakeholder participation integration. Cities. https://doi.org/10.1016/j.cities.2025.106214
47. Zhang X, Sun Y, Luan X (2023) Regulation flexibility and legitimacy building in governing intercity railways: the polymorphous role of the Chinese provincial government. J Urban Aff 1–16
48. Li Y, Zhang X, Xia C (2023) Towards a greening city: how does regional cooperation promote urban green space in the Guangdong-Hong Kong-Macau greater bay area? Urban For & Urban Green 86:128033
49. Zhang X, Shen J, Sun Y, Zhou C, Yang Y (2023) Formation of city regions from bottom-up initiatives: investigating coalitional developmentalism in the pearl river delta. Ann Am Assoc Geogr 113:700–716
50. Zhang X, Chen Y, Xu Y, Yue W, Tang C (2023) Does regional cooperation constrain urban sprawl? evidence from the Guangdong-Hong Kong-Macao greater bay area. Landsc Urban Plan 235:104742
51. Forsyth A, Oakes JM, Schmitz KH, Hearst M (2007) Does residential density increase walking and other physical activity? Urban Stud 44:679–697. https://doi.org/10.1080/00420980601184729
52. Long Y, Huang CC (2019) Does block size matter? The impact of urban design on economic vitality for Chinese cities. Environ Plan B Urban Anal City Sci 46:406–422
53. Habibian M, Hosseinzadeh A (2018) Walkability index across trip purposes. Sustain Cities Soc 42:216–225
54. Alshalalfah BW, Shalaby AS (2007) Case study: relationship of walk access distance to transit with service, travel, and personal characteristics. J Urban Plan Dev 133:114–118
55. Steadman P (2014) Density and built form: integrating "Spacemate" with the work of Martin and March. Environ Plan B Plan Des 41:341–358. https://doi.org/10.1068/b39141
56. Ye Y, Li D, Liu X (2018) how block density and typology affect urban vitality: an exploratory analysis in Shenzhen, China. Urban Geogr 39:631–652
57. Zhang X, Li Y, Zhou C, Luan X, Yuan F (2023) Rescaling of the land regime in the making of city-regions: a case study of china's pearl river delta. Urban Stud 60:483–500
58. Oliveira V, Medeiros V (2016) Morpho: combining morphological measures. Environ Plan B Plan Des 43:805–825

59. Saaty RW (1987) The analytic hierarchy process-what it is and how it is used. Math Model 9:161–176
60. Panchal S, Shrivastava AK (2021) A comparative study of frequency ratio, Shannon's entropy and analytic hierarchy process (AHP) models for landslide susceptibility assessment. ISPRS Int J Geo-Inf 10:603
61. Lee GKL, Chan EHW (2008) The analytic hierarchy process (AHP) approach for assessment of urban renewal proposals. Soc Indic Res 89:155–168. https://doi.org/10.1007/s11205-007-9228-x
62. Awad J, Jung C (2022) Extracting the planning elements for sustainable urban regeneration in Dubai with AHP (Analytic Hierarchy Process). Sustain Cities Soc 76:103496. https://doi.org/10.1016/j.scs.2021.103496
63. Zhao H, Shi J, Qi X, Wang X, Jia J (2016) Pyramid scene parsing network. IEEE Comput Soc 2882–2890
64. Su L, Chen W, Zhou Y, Fan L (2023) Exploring city image perception in social media big data through deep learning: a case study of Zhongshan city. Sustain 15:1–22. https://doi.org/10.3390/su15043311
65. Gu Q, Lu G (2023) Factors influencing the satisfaction level of college students in china: literature analysis based on grounded theory. Front Psychol 13:1–12. https://doi.org/10.3389/fpsyg.2022.1023420
66. Leung SM (2014) What makes the more habitable high-rise, high-density residential environment in the future of Hong Kong? Urban Des Int 1–26
67. Zheng D, Dong S, Lin C (2021) The necessity and control strategy of "medium density" in metropolis. Int Urban Plan 36:1–9
68. Huang J, Liang J, Yang M, Li Y (2022) Visual preference analysis and planning responses based on street view images: a case study of Gulangyu Island, China. Land 12:129
69. Zhao X, Chen J, Li J, Wang H, Zhang X, Yu F (2025) Unraveling the renewal priority of urban heritage communities via macro-micro dimensional assessment-a case study of Nanjing city, China. Sustain Cities Soc 124:106317
70. Song H, Chen J, Li P (2025) Decoding the cultural heritage tourism landscape and visitor crowding behavior from the multidimensional embodied perspective: insights from Chinese classical gardens. Tour Manag 110:105180. https://doi.org/10.1016/j.tourman.2025.105180
71. Zhao X, Yu F, Zhang X, Chen J, Li P (2025) Assessing Urban Renewal Efficiency via Multi-Source Data and DID-Based Comparison between Historical Districts. npj Herit Sci 13:389. https://doi.org/10.1038/s40494-025-01964-9.

Chapter 9
Densification Research by Design: The Pilot Case of Tuanjie Village in Kunshan

9.1 Introduction

The analysis of the urban regeneration development in China's new urbanization context, combined with the literature review presented in the previous chapters concludes that research should focus on socially sustainable regeneration, people-oriented decision-making process, and precise spatial improvement design [1, 2]. The previous empirical studies, focusing respectively on public consultation, decision-making and densification evaluation, suggest how a comprehensive working framework and adaptive and targeted design decisions can mitigate the contradiction between people and the environment.

J. Chen and P. Pellegrini, *Densification as a Design Tool for Sustainable Urban Regeneration*, Urban Sustainability, https://doi.org/10.1007/978-981-95-1653-7_9

The old community's regeneration should pursue not a simple renewal of building facades and minimal restructuring, but a holistic and integrated upgrade [3, 4]. Similarly it should not simply demolishing and rebuilding and expanding the urban area [3, 5], but increasing public infrastructure, central green space area, improving service quality, etc. [6, 7]. Densification can play an important role in facilitating this sustainable regeneration [4, 8].

At the policy decision-making level, densification is a long-term sustainable mechanism [9]. In terms of the spatial implementation aspect, as a way of guiding urban design, densification is a tool that can significantly improve spatial coordination and efficiency. It could be the response to many problems in the current old residential areas in China, such as incomplete infrastructure, insufficient functional support, lack of enough public space and parking lots, low building construction standards, etc.

In order to test the sustainable regeneration of the old communities in practice, this chapter present a research by design project conceived to upgrade the relevant spatial elements of an old community Taking Kunshan Tuanjie Village—namely, land use, infrastructure, residential buildings, and community culture—from a socially sustainable perspective [10, 11]. The project proposes a decision-making framework and urban design strategy, aiming at enhancing the functional synergy and the environmental quality based on residents' needs [12, 13]. The following research question is answered: what sustainable densification strategy best supports a sustainable community regeneration in high-density cities in China? What decision-making approach can best support a regeneration process with densification?

9.2 Sustainable Old Community Regeneration Design Review

9.2.1 Social Sustainability: Old Community Regeneration Orientation

Over the years the characters of a sustainable urban regeneration have evolved [14]. Sustainability is not only about the use of resources but also the approach to society and economy [15]. As early as 2003, the United Nations formally proposed a sustainable community plan to transform and update communities from the environmental, social, and economic aspects, related and interdependent [16]. Nowadays, social sustainability is a main topic in the contemporary old community regeneration research.

In terms of research method, the regeneration of old communities from the perspective of sustainability has expanded from traditional field surveys, questionnaire interviews, and photo interviews [17] to time–space positioning data, such as street view data, mobile signaling data, GPS positioning data [18, 19]. Through the combination of big and small data, residents' activities are cross verified and analyzed, which effectively enhances the data collection on community spaces and improves the objectivity and accuracy of cognitive space recognition.

In terms of research content, important aspects of the sustainable regeneration of old communities are how to enhance community participation through effective means [6, 8], improve the quality of life and enhance social participation, and strengthen the resonance between the community environment and community residents. The construction of complete communities with "humanization and precision" plays an important part in the community governance, because it strengthens community residents' cohesion and sense of belonging in coordinating multiple stakeholders [20].

9.2.2 'Sustainable Urban Form' Based Regeneration Design Modes

Urban design can be defined as a multi-disciplinary practice that shapes and manages the urban environment transforming the urban structure and space across multiple scales [21]. Beyond fulfilling functional needs, it enhances aesthetic quality and spatial experience while serving as a critical tool for sustainable community regeneration. However, despite its significance in urban renewal, urban regeneration design remains underemphasized in China.

Urban form control is an important part of urban design and the key to successful urban regeneration. Scholars identify several principles for sustainable urban form, including:

- Mixed land use (integrating residential, commercial, and office functions to reduce unnecessary travel and energy consumption) [3],
- High-density and compact development (optimizing land use, reducing traffic and transport emissions, and revitalizing existing urban areas) [6],
- Spatial and social and cultural diversity (in building typologies, cultural expression, and diversified social groups, as advocated by Jane Jacobs),
- Passive solar design (leveraging orientation, layout, and greenery to regulate the microclimate and minimize energy demand) [22],
- Green design (enhancing ecological balance and making cities as livable as suburbs) [7].

These elements collectively contribute to more sustainable, efficient, and vibrant urban environments.

9.2.3 Regeneration Implementation Framework Through Research by Design

Kunshan is within the administration of Suzhou in the Yangtze River Delta Economic Zone, about 50 km from Shanghai (Fig. 9.2), and was declared a model of China's economic development. It not only was praised for good economic development but it also indicates its willingness to be a pioneer of sustainable urban regeneration [23, 24]. Therefore, taking the competition launched by the local government for regenerating the old community of Tuanjie Village as a case study, this chapter discusses whether the urban design of old communities can coordinate multiple stakeholders and optimize the elements of community regeneration, that is community governance, community planning and community participation [25, 26]. The specific implementation steps are as follows (Fig. 9.1):

1. Design: Investigate and study the unique social environment and landscape ecology in and around the old community. On this basis, the current situation of community development, operation mode, regeneration management, etc. is analyzed, and the relationship between people and space is adjusted according to local conditions of community governance, community planning and residents' participation to form a preliminary plan for the sustainable regeneration of the old community.
2. Practice: Based on the design scheme, the four elements of regeneration, namely land use, infrastructure, residential buildings, and community building, are renovated, aiming to create an interactive atmosphere between residents and improve the sense of belonging and identity.
3. Research: Based on the plan, coordinate the stakeholders of the community regeneration, including the local government, community residents and other

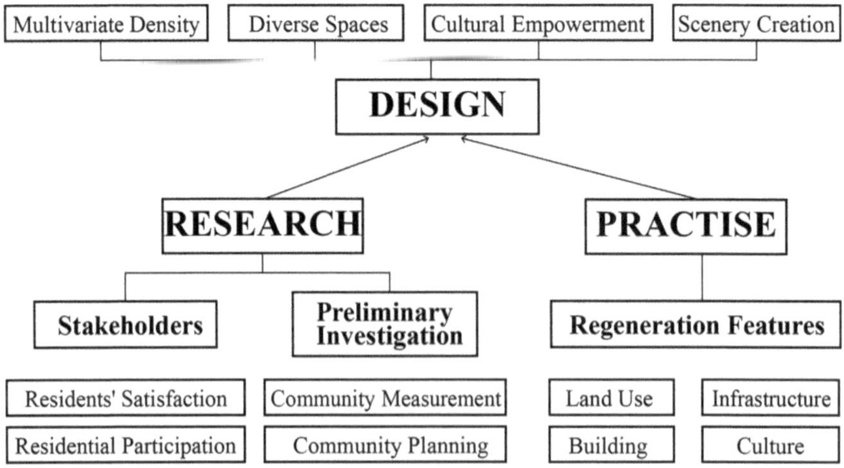

Fig. 9.1 Framework diagram of implementation steps

stakeholders having interests in the area, develop the corresponding regulatory detailed planning, and adjust the related design plan to enhance the satisfaction of the community residents.

9.3 Kunshan Tuanjie Village Regeneration

9.3.1 Preliminary Investigation

Since the reform and opening up, Kunshan has realized high-end industries in information technology under the umbrella of the open economic policy, which included the establishment of the National Economic and Technological Development Zone (ETDZ) in 1985 and the High tech Industrial Development Zone in 1994 [27]. The exponential GDP growth and population in a short time [28] made of Kunshan an exceptional economic and geographical case.

Kunshan is today an important part of the Yangtze River Delta Metropolitan Circle, a strategic area for China's economic development [29]. After the vigorous industrialization, the local government gradually pays attention to the quality of urban space [30]. According to the national guidelines, it indicates that "new urbanization" should not be limited to economic growth but also focus on urban regeneration to achieve high-quality urbanization [31]. In 2016, the Ministry of Housing and Urban–Rural Development awarded it the title of "National Ecological Garden City" and commended the ecological model of "water network, road network and green network" [32] (Fig. 9.2).

Tuanjie Village is a residential area located in the central part of Kunshan Chaoyang District, bordering Chaoyang West Road in the north and Hexing West Road in the south. The surrounding area is composed by residential neighborhoods

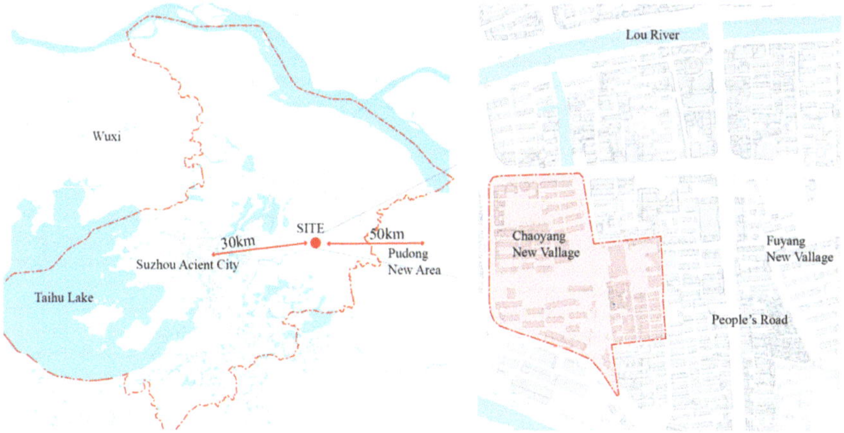

Fig. 9.2 Study are diagrams

Fig. 9.3 Collage of urban images in Tuanjie Village

and industrial plants (Fig. 9.3). The regeneration of the Village is prompted by the following problems, declared in the urban design competition brief promoted in 2022 by the Kunshan Municipal Government [33, 34].

1. Lack of car parking: the current demand, not originally planned, challenges the few available car parking spaces, resulting in disorderly organization and land use issues in and outside the community.
2. Lack of infrastructure: facilities such as fitness trails, elderly care, health care, power supply are missing, as well as waterproof and fire protection systems in the buildings.
3. Aging of buildings: low-cost construction, planned short-term life-span of the structures, no maintenance program means the buildings and the domestic pipelines, including sewage treatment systems, are decaying. In addition, elevators and heating systems are missing. The small dimension of some units has caused the limitation of residents' living functions, including a comfortable kitchen, bathroom with modern electrical appliances, etc.

   It is worth mentioning that, before 2001 the requirements for the "design service life" of new constructions were not specified. Since 2001, the Unified Standard for Reliability Design of Building Structures has updated the service life of building structures and components from 30 to 50–70 years [35]. At the beginning of the design process, based on the basic planning of Chaoyang District of Kunshan City, the buildings that can be demolished and the vacant in the area were defined (Figs. 9.4 and 9.5).
4. Excessive compactness. The intensive distribution of residential buildings has limited the privacy of individuals and reduced their living standards.
5. Low land use rate, a relatively thin cultural resource. In view of this situation, the research puts forward that the core elements of sustainable regeneration include: (1) the construction of urban open space and street scenes, (2) the improvement

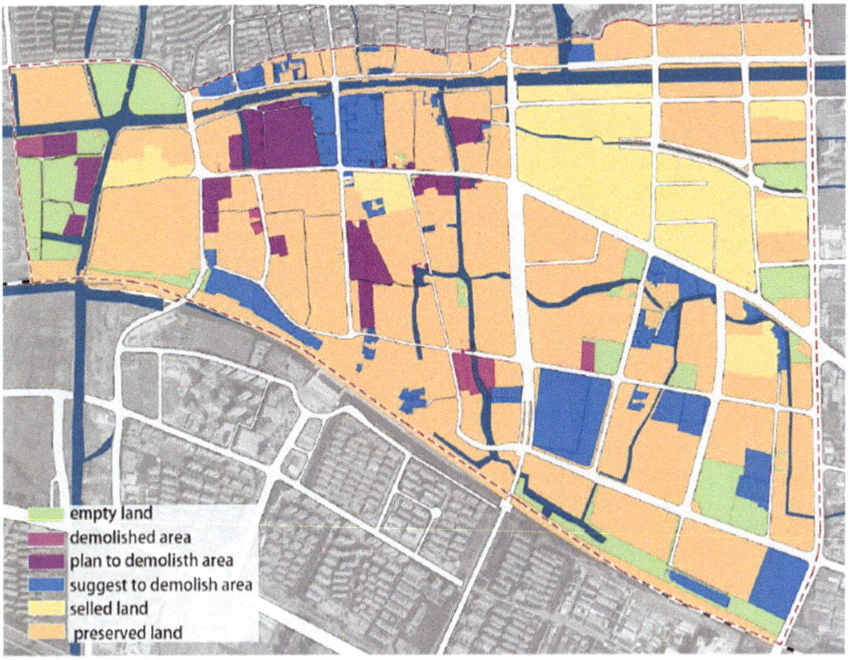

Fig. 9.4 Schematic diagram of overall planning

of green space landscape and supporting service facilities, and (3) the refinement and expression of historical and cultural connotation.

9.3.2 Urban Design System Planning of Old Communities from the Perspective of Sustainability

The project targeted the update of land use, infrastructure, residential buildings, and community construction to achieve socially sustainable regeneration (Fig. 9.6). The current human-cantered analytical model has shifted behaviour selection toward a focus on spatial search, making public layout and land-use structures increasingly human-oriented [36, 37].

1. Land use: the regeneration of land use in old communities is mainly aimed at dividing and re-planning community **public space functions** to meet the needs of community individuals, which is an important part of community regeneration [38]. The purpose is to solve a series of problems, such as make full use of idle land, disorder of property rights, disorder in functions, etc. [39]. Through sorting out the idle land and linking with multiple stakeholders, such as district committee offices, neighborhood offices, neighborhood committees, residents' representatives, and other stakeholders, a large amount of idle land was found to be usable.

Fig. 9.5 Schematic diagram of demolished or vacant buildings in Tuanjie Village

Fig. 9.6 Planning steps of urban design system

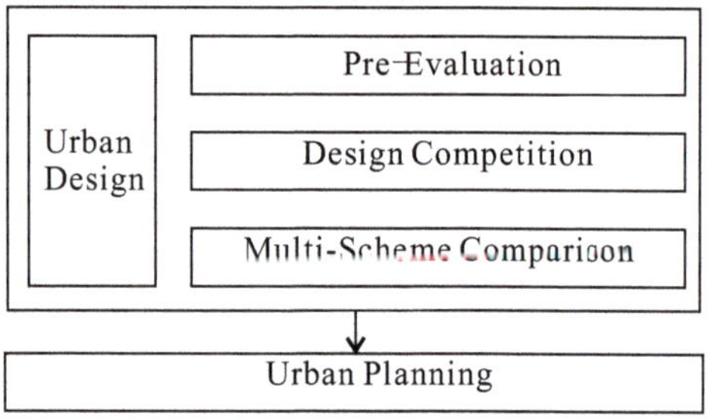

One part was destined to commercial use, the construction of barber shops, Yili shops, etc. This has accelerated the return of community regeneration funds, and diversified the economic development model through sharing funds and practical cooperation. The other part is used as a semi-public welfare activity to ease the "Conflict between rapid population growth and low land utilization" of the community. Through such actions, the community participation of community residents has also been strengthened (Fig. 9.7).

2. Infrastructure: community infrastructure is the basis for the quality of community public space and is also one of the important strategies to meet the needs of

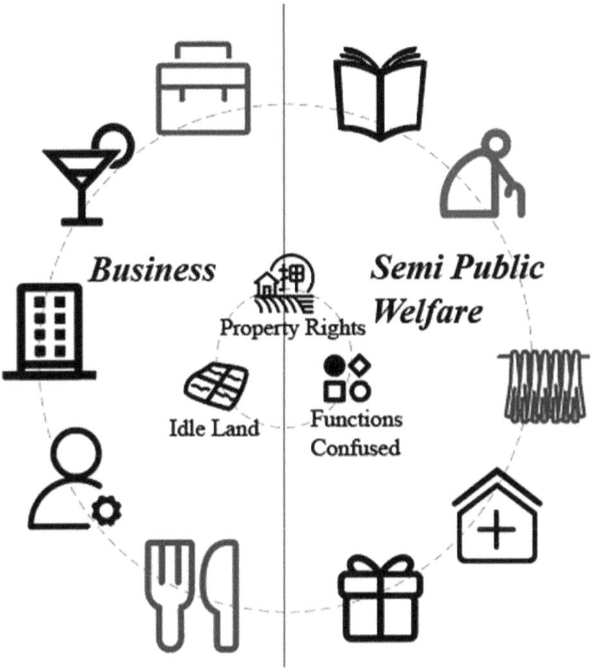

Fig. 9.7 Land use regeneration

community residents [40]. For example, the planning of community green helps to strengthen the interaction between community residents and see the improvement of the community environment [41]. Its main mean is to use the existing space resources, cooperate with the government, the market, and other stakeholders, and improve the community public space, including roof space, street space, residential building internal space (Fig. 9.8), community courtyard space and other public spaces, set up flexible infrastructure according to local conditions, and then build a small and flexible community public space. While utilizing existing resources, it is essential to promote the external opening of community infrastructure to the public, such as the normal open areas, and jointly manage with the venue operators and the community to provide convenient and optimized activity facilities for community residents. Revenues should be ensured to promote the community's sustainable development. Some communities have a generation through garbage classification, rainwater recycling and other means. We will also build charging facilities, public mopeds, and other means to meet the needs of smart energy conservation and recycling and create a low-carbon scene in the future.

3. Residential buildings: the regeneration of residential buildings in old communities is mainly to improve the basic standards of the residential living space.

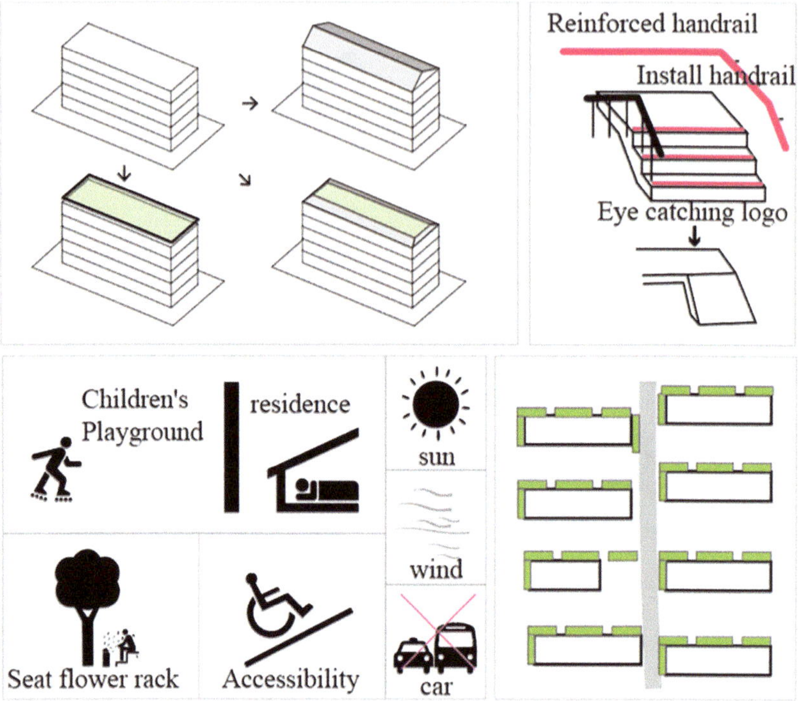

Fig. 9.8 Infrastructure improvements

Not only is the regeneration of residential buildings closely related to the relevant national policies, but also the regeneration of community residential buildings is related to the surrounding environmental system and ecosystem of the community.

Some scholars believe that the community residential buildings density is an important tool for predicting, controlling and evaluating land use [42]. Therefore, the reconstruction of old community residential buildings in China includes but is not limited to elevation renovation, roof reconstruction and pipeline maintenance. In recent years, the regeneration of the community has changed from the improvement of housing conditions dominated by units and individuals to the sustainable organic regeneration of community housing dominated by community participation. The transformation from decentralized partial demolition and reconstruction to organic community regeneration is the transformation of the regeneration subject and strategy and scale (Fig 9.9).

4. Community culture: community culture is the mobilization of people in the community and the process of completing self-organization, autonomy, and self-development. It has been usually divided it into five parts: human, cultural, local, production and landscape [43] (Fig. 9.10). Whether the community cultural circle is sound or not affects the degree of community participation and community

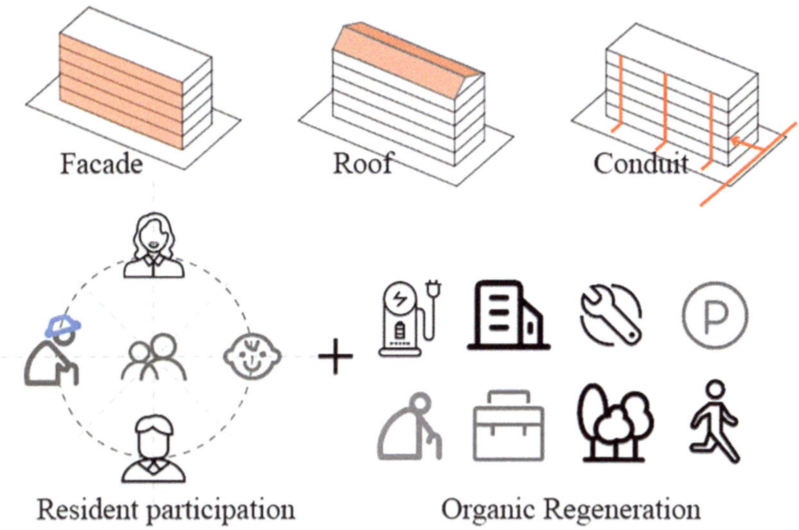

Fig. 9.9 Residential building regeneration

identity of community residents to a certain extent. Community-building mainly takes the measure of sharing private domains, refining community culture, and shaping iconic and lasting community brands to create a cultural aggregation and stimulate interaction and links between people. In this way, it could improve the enthusiasm of community residents to participate in various community affairs, thus breaking the restrictions of class circles, bringing together community residents with similar cultures, and achieving multi-level and multi-field communication and sharing.

9.4 Old Communities' Regeneration Implementation Strategies

Starting from the regeneration of community elements, namely, land use, infrastructure, residential buildings, and community culture (Fig. 9.11), the project aims to propose a scheme that can improve the quality of the existing Tuanjie Village Community urban environment.

9.4.1 Land Use: "Hybrid and Diversity"

The land use planning of Tuanjie Village streamlined the idle land inside the community, the grey space inside the building, the roof space, the public space, and other

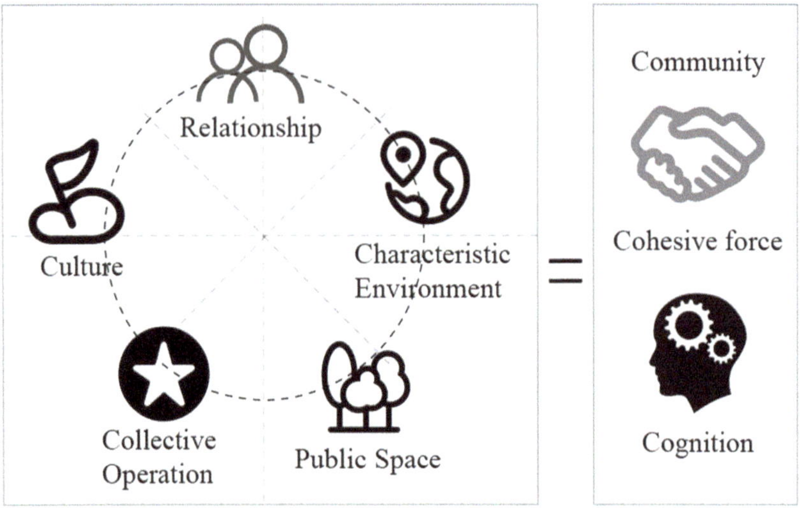

Fig. 9.10 Community-building methods

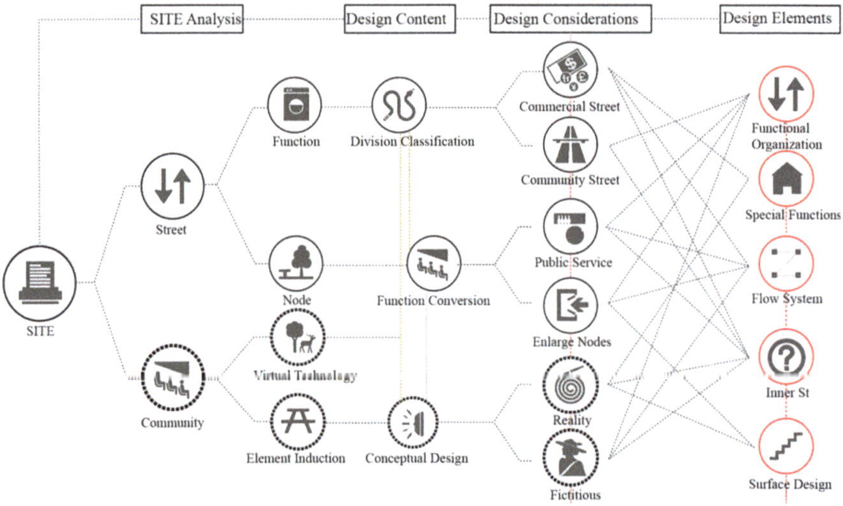

Fig. 9.11 Tuanjie community regeneration design process

spaces, regulate the property rights and make comprehensive use of the existing land [44]. The essence of land use regeneration is to ease the "contradiction between people and land", that is, that a high-density built environment struggles to meet the needs of community residents. Therefore, according to the interests of all stakeholders, the internal functional requirements of the community is reorganized. In combination with social life space relations, the external platform space is increased, vertical connections are introduced, and commercial space is integrated at the bottom of the buildings to facilitate community life.

Limited by the difference in the needs, the homogeneous land use, the low vitality of the internal area of the community, the mixed old and new buildings cannot meet the current needs of different stakeholders. Therefore, the "mixed and diversified" land use (Fig. 9.12) is proposed. Increasing the functions in the community, on the one hand, can meet the daily life needs of residents in the community. On the other hand, the diversified functions generated are intended to attract groups with young people as the main body and stimulate the site's vitality. For community residents, the northwest side of the site is combined with the loft to set up the bottom floor commerce. These commercial spaces are the main places for residents' daily consumption. The vegetable market in the northeast of Hexing Park and the elderly care community in the east of the plot are the supporting facilities of the community life circle, which significantly facilitates residents' life. As a new housing model, youth apartments and lofts attract many young people, alleviate the growing ageing of the community, and improve the diversity of community space and behavior. In addition, there are community gardens, roof gardens and other important ecological spaces in the community public green space network. While increasing the use of space ensures people's daily leisure and recreation needs, makes the most effective use of the urbanized resources, and maximizes the economic value of the land.

These spaces are designed using an array-based layout, emphasizing the homogeneity and sharing of resources, stagger layout, creating courtyard space, and sharing diversified resources. Part of the building height is lowered to form a descending space layout, create interspersed volume relations, enrich the social life space relations, increase the external platform space, establish vertical traffic connections, and integrate the commercial space at the bottom to facilitate life.

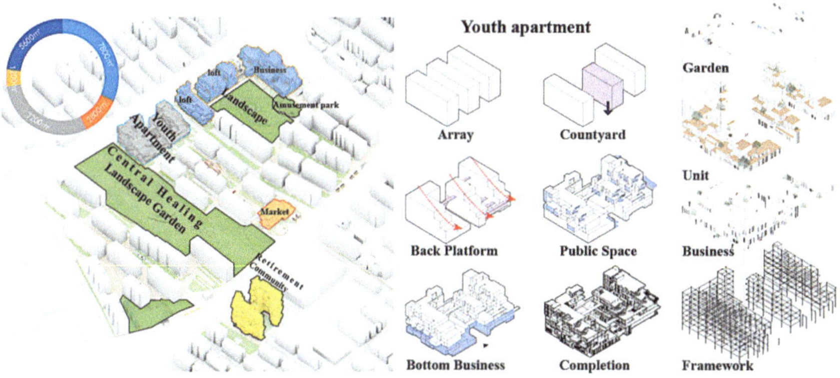

Fig. 9.12 "Mixed and diverse" land use and spatial design

9.4.2 *Infrastructure: "All-Age-Friendly"*

The upgrading of the community infrastructure in Tuanjie Village target a functional and comprehensive development. Based on the different ages within the community, the "green", "healthy", and "full age" upgrading of the infrastructure is proposed. The infrastructure within the community is mainly to coordinate the functional attributes between architectural spaces and increase the social characteristics of public spaces. The increased number and types of infrastructure includes the vegetable field (combined with the activity center), activity center, sports facilities, playground, bench, and canopy for residents to rest, etc.

The proposal of "green regeneration" is based on the survey of the urbanization life mode of surrounding communities. It is found that the traditional lifestyle of Tuanjie Village cannot adapt to the current policy demand of reducing carbon peak, and it is also difficult to adapt to the current green and low-carbon life mode. Therefore, the infrastructure of Tuanjie Village is sorted out based on "green regeneration". Health management facilities and low-carbon charging facilities are set in the node areas of public space to strengthen the system of green and low-carbon circular development. The "green regeneration" of infrastructure is not limited to the treatment of the sanitary environment but also strengthens the utilization of resources by setting rainwater return, water resource collection and wastewater recovery systems to optimize the allocation of energy resources to improve the utilization efficiency of resources. And the ecological environment within the community will be optimized with a focus on improving the quality and stability of the ecosystem to constantly improve the modernization level of the governance system and governance capacity in the ecological environment field, form a spatial pattern of resource conservation and environmental protection, and promote the continuous improvement of the ecological environment quality.

At the same time, community residents are paying attention to individual health. Therefore, the community's public space is excavated, including roof space, courtyard space and other public spaces, and the "all age-friendly" infrastructure is set flexibly and variously according to local conditions. It not only sets up fitness-supporting facilities to meet basic needs and enrich the basic "health" space but also sets up fitness trails for community residents to support their desire for a healthy life. While increasing the area of primary fitness space, an open and rational layout of space for all ages is created to achieve the infrastructure design with functions meeting the needs.

"Full age regeneration" is to realize the regeneration of community public transport infrastructure based on "healthy regeneration". The nodal space of the transportation system is positioned where the community center is. The interaction between people and the environment is strengthened through the bottom grey space, corridors, and overpasses between different building communities. For example, green space that can be used for activities and is not hindered by green decorative elements, a playground designed for children, benches and canopy that can be used for collecting

Fig. 9.13 Schematic diagram of "all age-friendly" infrastructure upgrading

and distributing, etc. The distributed node space design realizes "full age update" for all age groups of the community through scene connection (Fig. 9.13).

9.4.3 Community Buildings: "People-Oriented"

The regeneration of residential buildings under the sustainable perspective uses new technologies, new equipment, and new materials to produce renewable energy, including solar energy, wind energy, geothermal energy, etc. In the regeneration plan of residential buildings in Tuanjie Village, the green regeneration mode of "people-oriented first" is upheld.

According to the principle "people first" and cost control and in combination with the problems reflected by the community residents, such as the ageing of buildings and the lack of functional space, elevators are installed, the basic pipelines inside the building and the drainage system are repaired. Insulation panels for temperature protection inside the building are installed. In addition, a special sewage treatment and circulation system is set up. With the internal infrastructure as the basic unit, the system of external corridors is built to increase the public space, strengthen the relationship between community residents. In addition, different architectural styles and building types are combined for the different groups of community. The younger generation is mainly in apartments, while the elderly care home is built to achieve the integration of "treatment" and "maintenance".

The green regeneration of residential buildings is mainly divided into two aspects. On the one hand, the basic facade and roof of residential buildings are repaired with energy-saving and environment-friendly green materials to reduce the impact on the community ecology, and achieve the organic integration of community buildings and the lost environment. Green plants are introduced to create vertical greening, including basic climbing plants such as ivy, petunia, etc., or through roof gardens and green building methods, to optimize the ecological model of residential buildings, and then create a suitable living environment. On the other hand, through the introduction of new energy sources such as solar energy and natural gas, the original inefficient

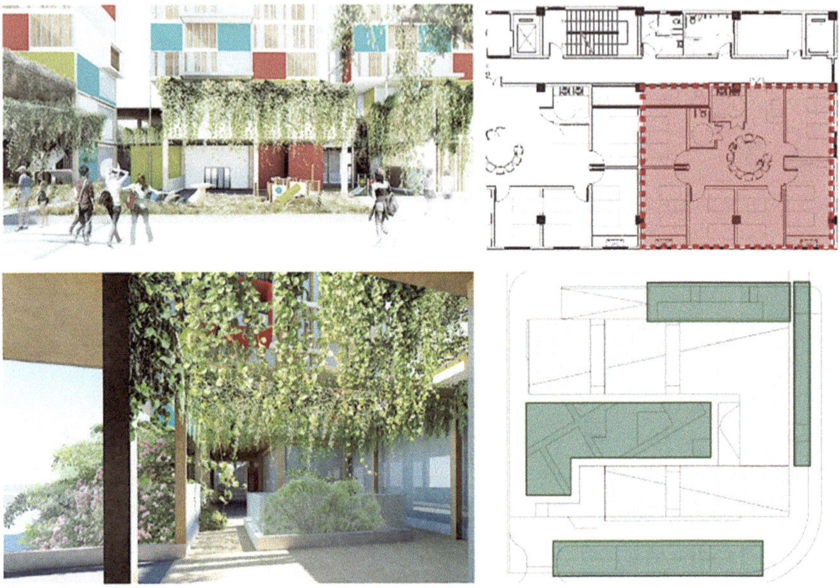

Fig. 9.14 Schematic diagram of residential building regeneration of "people-oriented priority"

and highly polluting resource utilization methods such as liquefied gas and carbon stoves will be banned to better integrate the living environment with nature and realize the sustainable regeneration mode of green development based on humanism (Fig. 9.14).

9.4.4 Community Culture: "Scenario Interaction"

Relevant research shows that there can be a positive emotional connection between individuals and the environment. The emotional connection similar to attachment formed by resonance with the place is conducive to triggering people's emotional memories, thus increasing the individual's access to space and utilization rate. Therefore, the updating mode of Tuanjie Village on community-building is to create a "situational interaction" space environment. Through a variety of social facilities and opportunity spaces, with "situational interaction" as the core, a space with the theme of "interaction" and "stay" is created, to realize the sense of belonging and attachment of community residents to the space.

Based on spaces with different attributes, the building methods of "scene interaction" are also different. For the leisure space in Tuanjie Village, the interactive space is created through different landscape elements, such as movable sunshades, chairs and other landscape furniture in the lawn, temporary sketches in the community garden, WIFI signals, power sockets, and spray painting and color painting of

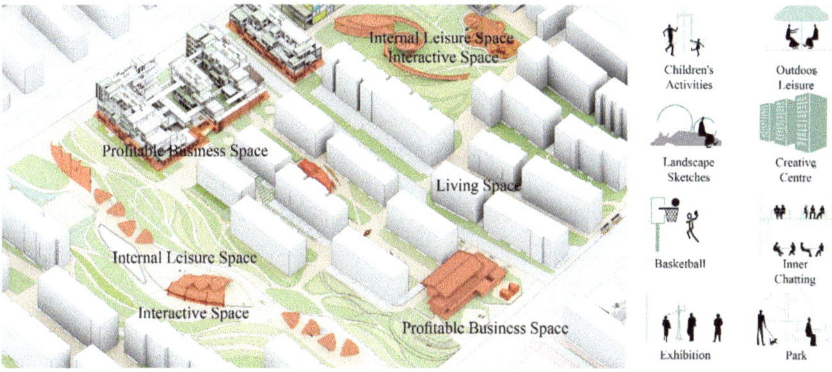

Fig. 9.15 Schematic diagram of community creation and updating of "Situational Interaction"

the space under the forest, so as to enhance the vitality of the space, and meet the basic social attributes of community residents. Unlike the public space for leisure, the commercial space with a profitable nature is used to create a cultural market based on the design of the community's food market and the community's internal culture. Through the standardization and regionalization of traditional stalls, the communication gap between various groups in the community is narrowed, and the stereotype of dirty, disorderly, and poor traditional market space is banned in places full of smoke and fire. At the same time, the "scene interaction" update of living and living space is mainly to blur the architectural boundaries through color painting and wall painting, and translate the space forms. The content of the wall painting is also diverse, both in line with the local culture of Kunqu's elegant charm and the young groups of starry innocence. The diverse content and convenient forms of participation significantly strengthen the community residents' sense of identity and belonging to the community culture.

It is worth mentioning that the public space within the community, in addition to adding landscape facilities to increase the "stay" of community residents, creates a demonstrative "new life", "new landmark" and "new conformity" cultural aggregation with unlimited mobility, space portability and mixed reality through streamline space design and the meaning of "winding water and flowing glass" (Fig. 9.15).

9.5 Research by Design

The regeneration project adopted a "research by design" method and actively carried out various investigations in the process of providing design schemes [45]. It defined the current problems, put forward corresponding concepts, proposed a first space design, verified the idea based on corresponding planning strategies and residents feedbacks, clarified space design means, integrated corresponding design techniques, redefined the problems to further understand user needs [45, 46].

9.5.1 Spatial Concept Development and Design Refinement

The proposed regeneration framework integrates sustainability principles with stakeholder needs across three dimensions: planning, governance, and participation. Four key spatial concepts guide the intervention: Mixed-use diversity for land optimization; Age-friendly infrastructure meeting universal needs; Human-centered design prioritizing resident well-being; Interactive environments blurring boundaries between users and spaces.

The concepts were refined through iterative testing, focusing on: Land use efficiency, Infrastructure modernization, Building upgrades, Community space activation. This process balanced technical requirements with humanistic values to create integrated solutions.

9.5.2 Design Improvement and Integration

From the perspective of scheme layout, the winning schemes of the competition have preserved (Fig. 9.16) the central garden, Hexing Garden, and built a complete public space system. In Scheme I–V, the air corridor is also involved in the construction of public space to expand the walking space of the site; The site facilities were improved through local demolition, reconstruction, and functional replacement (Table 9.1).

From the perspectives of spatial density and morphological continuity, the winning design schemes exhibit distinct strategic orientations. Scheme I demonstrates a high degree of contextual sensitivity by aligning its architectural texture with that of the surrounding old and new village buildings. While its form remains visually coherent

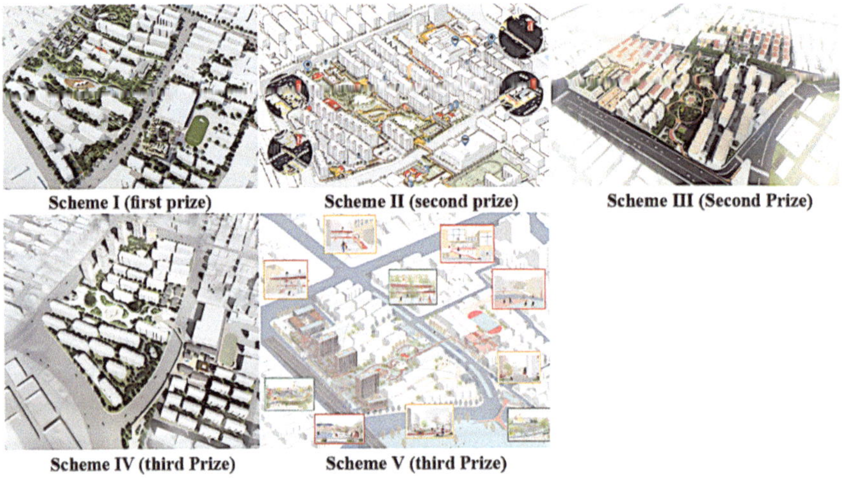

Fig. 9.16 Aerial views' comparison of winning schemes in urban design competition

Table 9.1 Comparison of different schemes in urban design competition

Plan	FAR	Building density	Characters
Present Situation	1.20	30%	It is a low-rise residential building built around 2000, with a low FAR (Floor Area Ratio)
Scheme I	1.21	40%	Continue the original texture, demolish the buildings and replace them with high-performance mixed interspersed volume buildings, with a large ecological space
Scheme II	1.30	33%	Continuation of the original space, with little change in spatial indicators
Scheme III	1.28	36%	
Scheme IV	1.66	37%	Break the original spatial pattern and implant high-rise building creation to increase the development possibility of the plot
Scheme V	1.71	31%	

within the existing environment, its spatial organization emphasizes both functional diversity and density. The design employs smaller modular units to achieve flexible spatial combinations, thereby enhancing efficiency in land use. In addition, the plan reserves substantial public space—most notably a large public area on the north side of Tuanjie Village—while transforming demolished and rebuilt structures into multi-generational leisure and activity hubs.

One of the key strengths of Scheme I lies in its ability to break away from the homogeneity of traditional space models. It proposes a spatial paradigm that seeks to create a demonstrative "new life," "new symbol," and "new landmark" while fostering a sense of community unity. Importantly, the project retains the original urban scale and spatial grain, maintaining the distinctive rhythm of the existing village fabric. Through the conceptual metaphor of "vine and tree," the scheme envisions a public space network that expands organically in multiple directions. This approach not only fosters spatial continuity but also strengthens the functional interconnections within the site, allowing public space to become a catalyst for both architectural expression and environmental activation.

In contrast, Schemes IV and V adopt a more segmented spatial strategy. By introducing high-rise buildings on the northwest side, these schemes aim to increase site compactness and accommodate additional functions. However, although the central and eastern zones preserve the original spatial texture, the vertical expansion in selected areas results in a partial disruption of the traditional spatial character. While the increased building volume may enhance functional intensity, it simultaneously challenges the coherence of the local urban identity, highlighting the tension between intensification and heritage continuity.

9.6 Problem Redefinition

Kunshan Municipal Government has revised the submitted design schemes in combination with the feedback from various stakeholders, including the District Committee Office, neighborhood offices, neighborhood committees, resident representatives, and other stakeholders, and form the relevant upper-level planning (Fig. 9.17). Overall, stakeholders were satisfied with the design schemes, and approved the space update mode of "mixed and diverse", "all age friendly", "people-oriented first" and "situational interaction". They also appreciated the comprehensive shaping of the "continuity" of the community center landscape, such as the healing landscape garden. The "continuity" landscape comes from the landscape preference matrix proposed that is, the "continuity" and "consistency" landscape is conducive to the recovery of residents' attention and the coordination of the relationship between man and nature. On this basis, Kunshan Municipal Government strengthens the connection with the central landscape through the corridors between buildings, and further promotes the connection between community residents and the landscape environment.

Fig. 9.17 Aerial view of regulatory plan by local government. (*Source* Kunshan Municipal People's Government)

9.7 Conclusion

The sustainable regeneration of old communities should not only be a micro transformation of the built environment, but consider of all aspects of the existing community, including the ways of community residents' participation.

This research takes urban design as a breakthrough mean to coordinate the solution of the existing planning problems in a multi stakeholder old community, Kunshan Tuanjie Village. Four types of regeneration elements of the old communities are tackled by the project: land use, infrastructure, residential buildings, and community culture.

The main problem of the old community is that the existing environment does not meet the growing physical spatial needs of residents, and the existing cultural environment does not reflect the needs of community residents, including a series of community problems such as difficult parking, old facilities, and small space. Therefore, in the design process, with the goal of densification, four types of regeneration modes, namely "hybrid and diverse", "all age friendly", "people-oriented" and "scenario interaction" are proposed for land use, infrastructure, residential buildings, and community culture respectively [47]. The "continuity" landscape design concept is used to coordinate the four types of regeneration models, create the community's core landscape environment, such as the community's central healing landscape garden, leisure and lively garden, rainwater garden landscape, etc., so as to coordinate community planning issues, community residents' participation and stakeholders in the process of community regeneration. Through the feedback on the design, it is found that the concept of "continuity" is widely recognized by all stakeholders. They believe that the construction of "continuity" should not be limited to the construction of community landscape environment. Spatial corridors and linkages are used to connect some buildings to strengthen the internal connection of the community and the interaction between residents within the community, thus forming the final plan [48, 49].

To sum up, the urban design with densification strategy for the regeneration of old communities is conducive to optimizing the spatial organization relationship of communities. The densification as a tool is used to improve the design of four types of regeneration elements of community land use, infrastructure, residential buildings, and community-building from the three levels of residents' participation, community planning and community governance, and to release the contradiction between the status quo and community residents' needs. It is also conducive to ensuring the positive interaction between community residents and the environment, strengthening community participation, thereby strengthening the sense of belonging and centripetal force of community residents, and achieving sustainable community regeneration. "Research by design" is used to coordinate multiple stakeholders, adjust, and improve the design scheme, and redefine practical problems. This design practice research practice mode is conducive to enhancing the recognition of the scheme, promoting the public participation of community residents and other stakeholders, and thus realizing the sustainable regeneration.

References

1. Zhang X, Lu Y, Xu Y, Zhou C, Zou Y (2024) Governing regional inequality through regional cooperation? A case study of the Guangdong-Hong Kong-Macau greater bay area. Appl Geogr 162:103135. https://doi.org/10.1016/j.apgeog.2023.103135
2. Zhang X, Zhou Z, Song Y (2004) Contested geopolitics of distribution in the city-region-building: promoting enclave industrial parks in the pearl river delta, China. Trans Inst Br Geogr e12697
3. Chen J, Ren K, Li P, Wang H, Zhou P (2024) Toward effective urban regeneration post-covid-19: urban vitality assessment to evaluate people preferences and place settings integrating IBSNS and POI. Environ Dev Sustain. https://doi.org/10.1007/s10668-024-05378-8
4. Zhao M, Liu N, Chen J, Wang D, Li P, Yang D, Zhou P (2024) Navigating post-COVID-19 social–spatial inequity: unravelling the nexus between community conditions, social perception, and spatial differentiation. Land 13
5. Hu J, Chen J, Li P, Yan J, Wang H (2024) Systematic review of socially sustainable and community regeneration: research traits, focal points, and future trajectories. Build 14:881. https://doi.org/10.3390/buildings14040881
6. Zhang X, Chen J, Wang H, Yang D (2025) From policy synergy to equitable greenspace: unveiling the multifaceted effects of regional cooperation upon urban greenspace exposure inequality in china's Megacity-Regions. Appl Geogr 174:103472. https://doi.org/10.1016/j.apgeog.2024.103472
7. Chen J, Li P, Wang H (2025) Assessment of influential factors on commute and life satisfaction in a historic campus-adjacent environment: evidence from a comparison study of twin cities. J Urban Plan Dev 151:4024065. https://doi.org/10.1061/JUPDDM.UPENG-5171
8. Chen J, Zhao X, Wang H, Yan J, Yang D, Xie K (2024) Portraying heritage corridor dynamics and cultivating conservation strategies based on environment spatial model: an integration of multi-source data and image semantic segmentation. Herit Sci 12:419. https://doi.org/10.1186/s40494-024-01497-7
9. Wu F (2022) Creating Chinese urbanism: urban revolution and governance changes. UCL Press, ISBN 1800083335
10. Chen J, Li P, Wang H (2024) MSPA-based collaborative evaluation and planning governance of territory and natural environment in metropolitan agglomerations—a case study of egiddz in Yangtze river delta. J Resour Ecol 15:1263–1273. https://doi.org/10.5814/j.issn.1674-764x.2024.05.014
11. Pan L, Gan W, Chen J, Xian X (2024) Ecological planning for mitigating urban heat based on thermal network simulation: a case study of Guangzhou-Foshan metropolitan area. J Chin Urban For 22:10–19. https://doi.org/10.12169/zgcsly.2024.06.26.0004
12. Chen J, Gan W, Liu N, Li P, Wang H, Zhao X, Yang D (2024) Community quality evaluation for socially sustainable regeneration: a study using multi-sourced geospatial data and AI-based image semantic segmentation. ISPRS Int J Geo-Inf 13(5):167. https://doi.org/10.3390/ijgi13050167
13. Chen J, Pellegrini P, Yang Z, Wang H (2023) Strategies for sustainable urban renewal: community-scale GIS-based analysis for densification decision making. Sustain 15:7901
14. Yun Y, Hu J, Ren L (2020) Philosophical thinking on sustainable urban form. Urban Plan Forum 32–40
15. McDonald S, Malys N, Maliene V (2009) Urban regeneration for sustainable communities: a case study. Technol Econ Dev Econ 15:49–59
16. Dale A, Onyx J (2010) A dynamic balance: social capital and sustainable community development. UBC Press, ISBN 0774859776
17. Zhai B, Ng MK (2013) Urban regeneration and social capital in china: a case study of the drum tower Muslim district in Xi'an. Cities 35:14–25
18. Zhao X, Chen J, Li J, Wang H, Zhang X, Yu F (2025) Unraveling the renewal priority of urban heritage communities via macro-micro dimensional assessment-a case study of Nanjing city, China. Sustain Cities Soc 124:106317

19. Song H, Chen J, Li P (2025) Decoding the cultural heritage tourism landscape and visitor crowding behavior from the multidimensional embodied perspective: insights from Chinese classical gardens. Tour Manag 110:105180. https://doi.org/10.1016/j.tourman.2025.105180
20. Ding F, Ren Y (2021) Analysis of public space in historic districts of historic cities based on community governance. Convert 80–87
21. Madanipour A (1997) Ambiguities of urban design. Town Plan Rev 363–383
22. Yang N, Chen J, Ban L, Li P, Wang H (2024) Pre-planning and post-evaluation approaches to sustainable vernacular architectural practice: a research-by-design study to building renovation in Shangri-La's Shanpian House, China. Sustain 16:9568
23. Ma G, Paola P, Wu H, Han H, Wang D, Chen J (2023) Impact of land-use mixing on the vitality of urban parks: evidence from big data analysis in Suzhou, Yangtze river delta region, China. J Urban Plan Dev 149:4023045. https://doi.org/10.1061/JUPDDM.UPENG-4334
24. Chen J, Wang H, Yang Z, Li P, Ma G, Zhao X (2023) Comparative spatial vitality evaluation of traditional settlements based on suf: taking Anren ancient town's urban design as an example. Sustain 15:8178
25. Pan L, Gan W, Chen J, Ren K (2023) An integrated model for constructing urban ecological networks and identifying the ecological protection priority: a case study of Wujiang District, Suzhou. Sustain 15:4487
26. Chen J, Tian W, Xu K, Pellegrini P (2022) Testing small-scale vitality measurement based on 5D model assessment with multi-source data: a resettlement community case in Suzhou. ISPRS Int J Geo-Inf 11:626. https://doi.org/10.3390/ijgi11120626
27. Gao J, Wei YD, Chen W, Yenneti K (2015) Urban land expansion and structural change in the Yangtze river delta, China. Sustain 7:10281–10307. https://doi.org/10.3390/su70810281
28. Kunshan Statistics Bureau (2019) Kunshan national economic and social development statistics bulletin
29. Xinhua news agency outline of the Yangtze river delta regional integration development plan
30. Chen J, Pellegrini P, Wang H (2022) Comparative residents' satisfaction evaluation for socially sustainable regeneration—the case of two high-density communities in Suzhou. Land 11:1483. https://doi.org/10.3390/land11091483
31. Xinhua News Agency Press Conference on the 'National New Urbanization Plan (2014–2020)'
32. Chinese Association of Park Notice of National Ecological Garden City, Garden City, County and Town in 2015
33. Zhang X, Li Y, Zhou C, Luan X, Yuan F (2023) Rescaling of the land regime in the making of city-regions: a case study of China's pearl river delta. Urban Stud 60:483–500
34. Zhang X, Chen Y, Xu Y, Yue W, Tang C (2023) Does regional cooperation constrain urban sprawl? evidence from the Guangdong-Hong Kong-Macao greater bay area. Landsc Urban Plan 235:104742
35. Ministry of housing and urban-rural development of China unified standard for reliability design of building structures
36. Zhang X, Shen J, Sun Y, Zhou C, Yang Y (2023) Formation of city regions from bottom-up initiatives: investigating coalitional developmentalism in the pearl river delta. Ann Am Assoc Geogr 113:700–716
37. Li Y, Zhang X, Xia C (2023) Towards a greening city: how does regional cooperation promote urban green space in the Guangdong-Hong Kong-Macau greater bay area? Urban For & Urban Green 86:128033
38. Souza L, Bueno C (2022) City information modelling as a support decision tool for planning and management of cities: a systematic literature review and bibliometric analysis. Build Environ 207:108403. https://doi.org/10.1016/j.buildenv.2021.108403
39. Wahidah AN, Martono DN (2022) Supriatna land use sustainability to mitigate potential land slide in Ciletuh watershed, Sukabumi, Indonesia. IOP Conf Ser Earth Environ Sci 950. https://doi.org/10.1088/1755-1315/950/1/012006
40. Brown J, Barber A (2012) Social infrastructure and sustainable urban communities. In: Proceedings of the proceedings of the institution of civil engineers-engineering sustainability, vol 165. Thomas Telford Ltd, pp 99–110

41. Mell IC (2009) Can green infrastructure promote urban sustainability? In: Proceedings of the proceedings of the institution of civil engineers-engineering sustainability, vol 162. Thomas Telford Ltd, pp 23–34.
42. Boyko CT, Cooper R (2012) High dwelling density as sustainability solution in Lancaster. Proc Inst Civ Eng Eng Sustain 165:81–88. https://doi.org/10.1680/ensu.2012.165.1.81
43. Wu Y, Ren L, Li K (2019) Sustainability-oriented social innovation design: the value of co-creation. In: Proceedings of the 2019 IEEE Eurasia Conference on Biomedical Engineering, Healthcare and Sustainability (ECBIOS), IEEE, pp 162–165
44. Zhang X, Sun Y, Luan X (2023) Regulation flexibility and legitimacy building in governing intercity railways: the polymorphous role of the Chinese Provincial Government. J Urban Aff 1–16
45. Mosely G, Wright N, Wrigley C (2018) Facilitating design thinking: a comparison of design expertise. Think Ski Creat 27:177–189
46. Micheli P, Wilner SJS, Bhatti SH, Mura M, Beverland MB (2019) Doing design thinking: conceptual review, synthesis, and research agenda. J Prod Innov Manag 36:124–148
47. Zhao X, Yu F, Zhang X, Chen J, Li P (2025) Assessing Urban Renewal Efficiency via Multi-Source Data and DID-Based Comparison between Historical Districts. npj Herit Sci 13:389. https://doi.org/10.1038/s40494-025-01964-9.
48. Ni H, Chen J, Li P (2025) Regeneration Efficiency Assessment and Predictive Comparison of Government-Led and Market-Driven Models in Historic Districts Via DID and XGBoost. Appl Spat Anal Policy 18:147. https://doi.org/10.1007/s12061-025-09749-0.
49. Pellegrini P, Chen J (2025) The Regeneration Path Not Taken Further: An Experiment in Urban Densification and State Entrepreneurialism in a Resettlement Neighborhood in Suzhou. China Urban Stud 00420980251365188. https://doi.org/10.1177/00420980251365188.

Chapter 10
Densification via Non-demolition

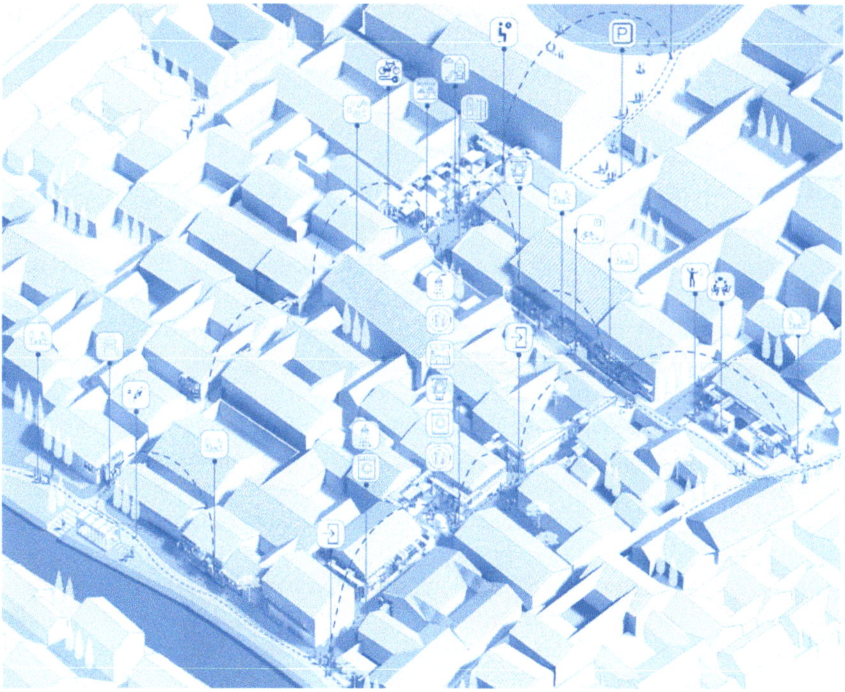

10.1 Introduction

As urbanization accelerates, traditional expansion methods are no longer sufficient to meet the demands of modern cities, especially given the increasing scarcity of land and the pressing need for environmental sustainability [1, 2]. Densification, as an effective urban design strategy, has become widely adopted in urban renewal

J. Chen and P. Pellegrini, *Densification as a Design Tool for Sustainable Urban Regeneration*, Urban Sustainability, https://doi.org/10.1007/978-981-95-1653-7_10

and regeneration projects, particularly in the revitalization of aging communities [3, 4]. Densification not only maximizes land use efficiency but also promotes the concentrated utilization of resources, helping to mitigate the negative environmental impact of urban sprawl [5, 6].

Traditional urban renewal models often emphasize demolition and reconstruction. This approach, however, is resource-intensive, energy-consuming, and potentially disruptive to social structures and cultural heritage [7, 8]. In recent years, the "non-demolition" approach, which focuses on adding new structures rather than tearing down existing buildings, has gained traction. This method preserves the historical essence of buildings while enabling densification without disrupting the established community structure [9, 10]. By balancing sustainable development with societal needs, this approach demonstrates significant potential in urban regeneration.

This chapter explores the application of non-demolition densification in the renewal of aging communities through specific case studies. It analyzes its role in urban regeneration and highlights both the advantages and challenges of this design approach in practice [11, 12]. By examining typical examples in detail, this chapter aims to provide fresh insights for urban regeneration, promoting a more sustainable and human-centered approach to urban development [13, 14].

10.2 Densification Case Study in Hongkong

This section analyzes the "non-demolition infill" densification design approach through a residential complex in Hong Kong. By adding new structures on existing buildings, this project maximizes the use of limited urban space and explores a high-density approach to urban regeneration [15, 16]. Located on Hong Kong Island, the project addresses the pressing issue of land scarcity.

Hong Kong faces significant housing challenges, with residents often dealing with cramped living spaces and poor quality of life [17, 18]. As land resources continue to dwindle, efficient densification strategies on existing land have become the project's core focus. With most residential areas on Hong Kong Island consisting of high-rise buildings and land availability increasingly limited, the project aims to alleviate land shortages through an innovative "aerial housing" construction model [19, 20].

The design team proposed a solution where residential units are suspended in the air, supported by a curved grid structure. By avoiding demolition and building on the existing infrastructure, the project utilizes "suspended structure" technology to transform 20-foot international standard shipping containers into residential units, which are then anchored to the main steel frame. This innovative approach allows the team to achieve high-density living spaces on available land. Figures 10.1 and 10.2 illustrates the concept of this suspended structure, showing how residential units are positioned in the air, minimizing land use while preserving existing infrastructure. This method not only improves land use efficiency but also enhances the appearance and functionality of the building, fostering a vibrant community.

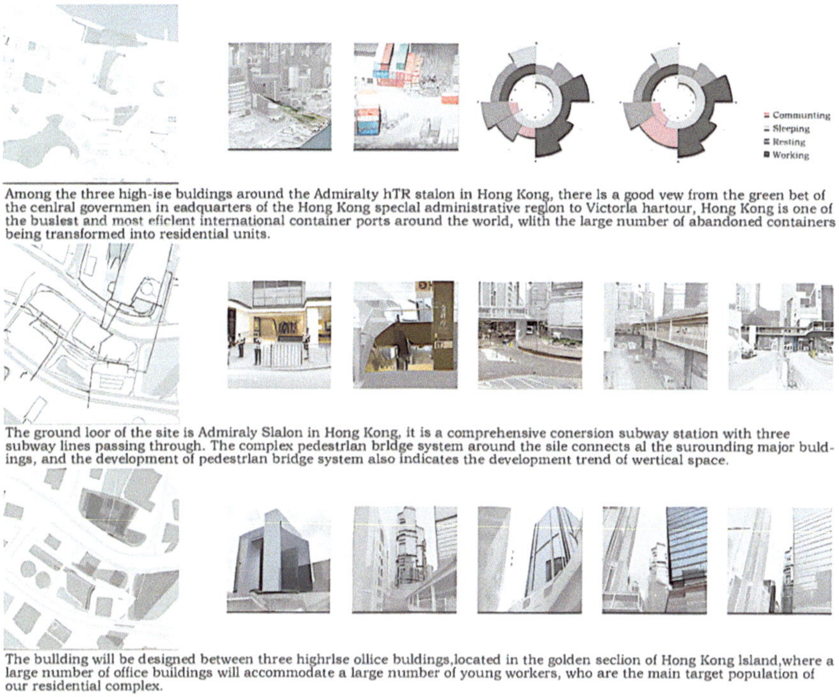

Fig. 10.1 Study area and question analysis

To cater to the needs of young residents, the design team carefully planned the layout of residential units and public spaces. Each unit is equipped with compact yet functional kitchens, bedrooms, and workspaces, while the public areas feature recreational platforms, sports facilities, and community spaces. Figure 10.3 highlights the spatial layout, emphasizing how public platforms enhance interaction and socialization among residents. Public kitchens and shared spaces, such as photography studios and fitness areas, are specifically designed for young people, creating a multifunctional social environment. In addition, multiple external activity spaces were designed, transforming containers into leisure and exercise areas to strengthen residents' sense of community.

From a sustainability perspective, the project prioritizes green building and energy-saving design. By incorporating passive solar design and green roofs, it enhances both the building's energy efficiency and the residents' living environment. Figure 10.4 shows the green roof plan, where the vegetation not only beautifies the environment but also regulates the microclimate, reducing energy consumption. The project also features an energy-efficient water management system that processes and distributes both clean water and wastewater, improving resource utilization and minimizing reliance on external resources.

Fig. 10.2 Design and renderings

Considering the project's location in a busy commercial district, the design team focused on optimizing traffic flow. Pedestrian pathways surrounding the two public centers ensure safe and convenient movement for residents. The project also enhances connections with surrounding commercial and office areas, allowing residents easy access to various services and amenities. Figure 10.5 illustrates the section design,

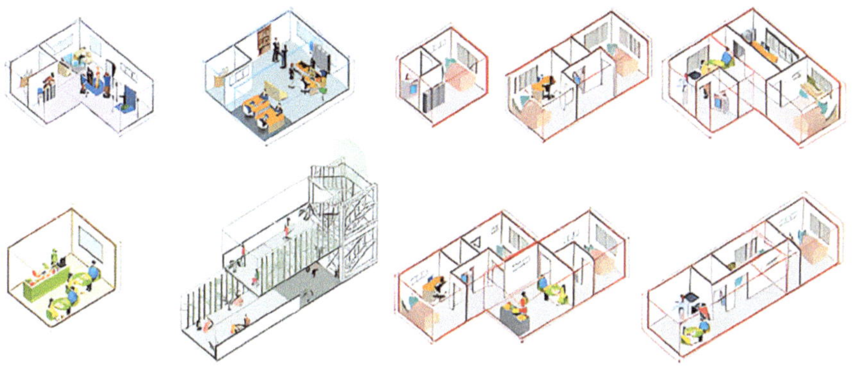

Fig. 10.3 Inner spatial design analysis

Fig. 10.4 Design sustainable analysis

highlighting the pedestrian and vehicle separation system that ensures a safer and more comfortable living experience in a bustling urban environment.

In summary, this case study demonstrates how the "non-demolition infill" densification design approach can innovate and enhance existing communities while maintaining their original characteristics. This design method not only increases land use efficiency but also offers new insights into urban regeneration, particularly in cities facing land scarcity.

10.3 Densification Case Study in Pingjiang Road District, Suzhou

The Suzhou Pingjiang case study presents another form of densification and infill, achieved through vertical expansion of existing buildings to meet the growing urban housing demand. Located in a highly dense urban area, the project aims to

Fig. 10.5 Design transport analysis

enhance land use efficiency without demolishing the original buildings, preserving the community's structure and cultural identity.

As urban populations rapidly grow, land resources become increasingly scarce. Traditional urban renewal often relies on demolition and reconstruction, which not only wastes resources but may also disrupt social structures. Vertical expansion presents an ideal solution. This project adds multiple residential and commercial levels atop an existing low-rise commercial building, meeting the spatial needs of a dense city while optimizing space usage (Fig. 10.6). The site was originally a low-rise commercial building surrounded by dense, functionally limited structures. To address land scarcity and the needs of residents in the commercial district, the design team proposed a vertical expansion. By adding several floors for residential and office spaces, the project effectively increases the building's capacity while maintaining the existing infrastructure and avoiding the environmental impact of demolition.

In this project, vertical expansion increases building height, accommodating more residents and commercial space. The design uses sustainable and cost-effective structural solutions, employing lightweight materials and steel supports to ensure the safety and stability of the added floors. The original structure is reinforced, and the addition of elevators, redesigned staircases, and improved seismic resistance extends the building's lifespan. Figure 10.7 shows the overall aerial rendering and facade design of the expanded portion, where new residential floors seamlessly integrate with the existing building, maintaining aesthetic harmony.

The design particularly emphasizes integration with the surrounding environment, ensuring a smooth transition between old and new structures through simple facades and appropriate landscaping. The project also takes into account the need for functional diversity. In addition to vertical expansion, the design team created shared public spaces between buildings, such as recreational plazas, green areas, and multifunctional activity rooms. These spaces allow residents to enjoy private comfort while using shared amenities to foster community interaction. Residential units were

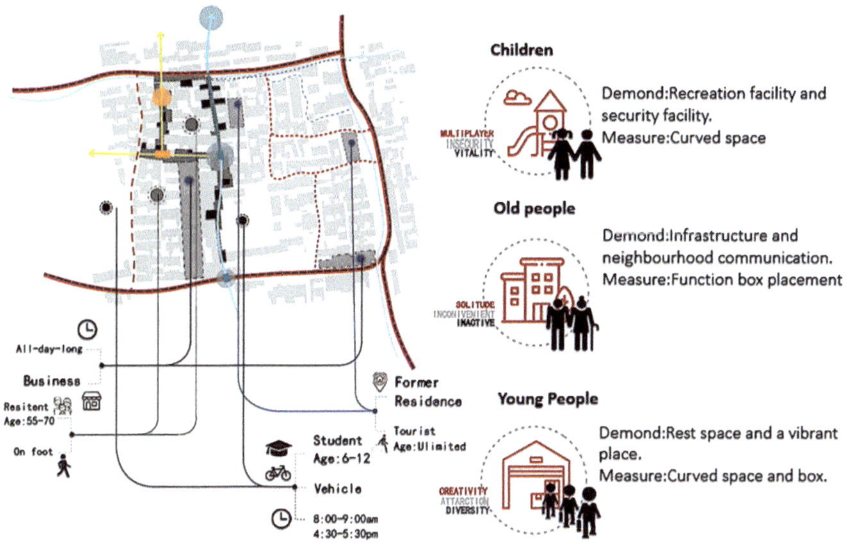

Fig. 10.6 Study area and analysis

planned according to the needs of target residents, offering a range of layouts, including studio apartments, one-bedroom, and two-bedroom units. Figure 10.8 shows the floor plans of the residential levels, highlighting how the design maximizes space utilization with efficient circulation and furniture arrangements.

To ensure sustainability, the project emphasizes green building principles, especially in energy efficiency and environmental impact. The newly added floors feature low-energy windows and green roof designs, utilizing natural light and ventilation to reduce air conditioning use and improve energy efficiency. Additionally, solar panels and rainwater harvesting systems enhance the building's environmental performance. Figure 10.9 shows the green roof and solar panel layout, improving both the building's aesthetic and its sustainability by reducing reliance on external energy sources. This green design minimizes environmental impact while supporting high-density.

Overall, this case study demonstrates how vertical expansion on existing buildings not only increases land use efficiency but also improves residents' quality of life through smart functional design and green principles. This method offers valuable insights for regenerating and updating old buildings in urban environments, proving the feasibility and necessity of densification strategies in modern cities.

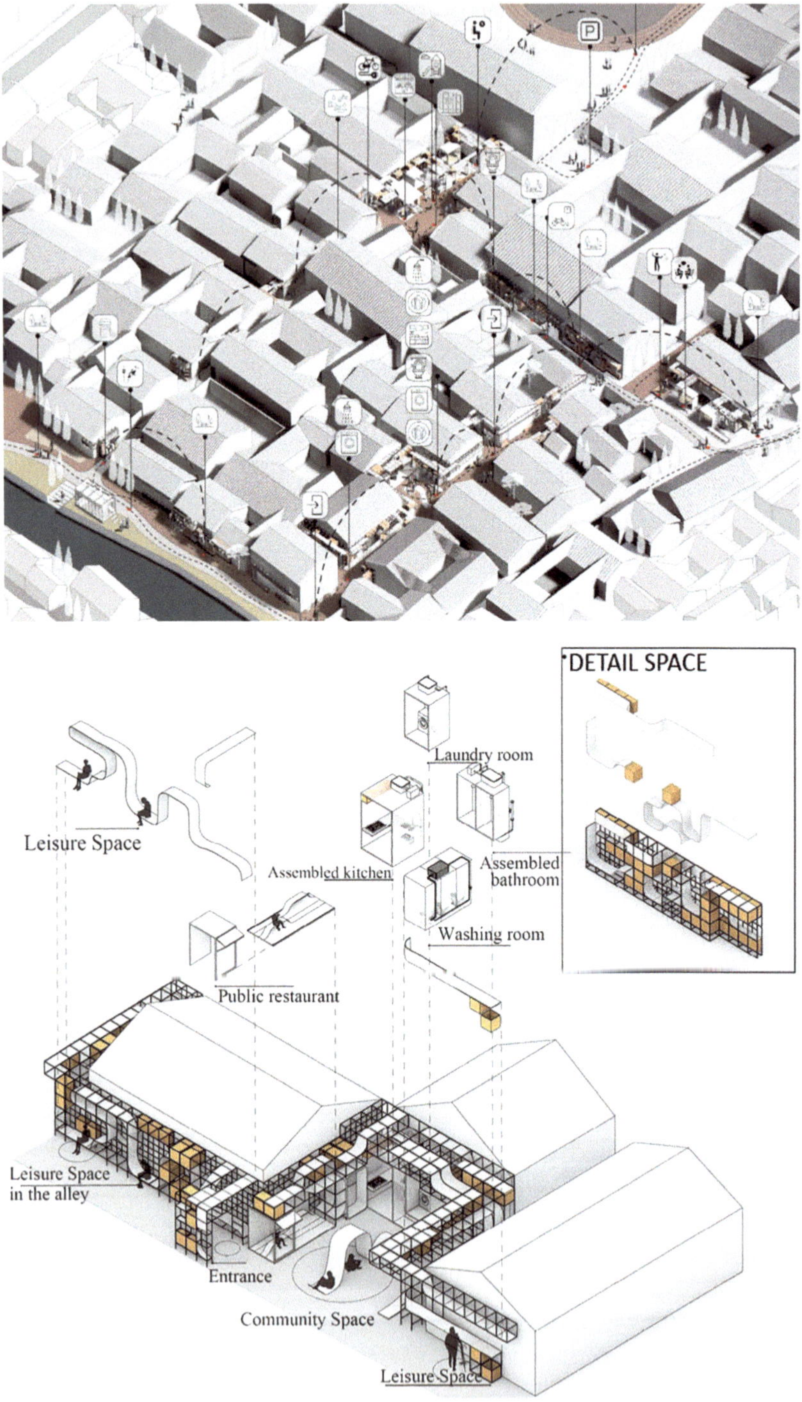

Fig. 10.7 Overall aerial rendering and facade analysis

Fig. 10.8 Design functional diversity renderings

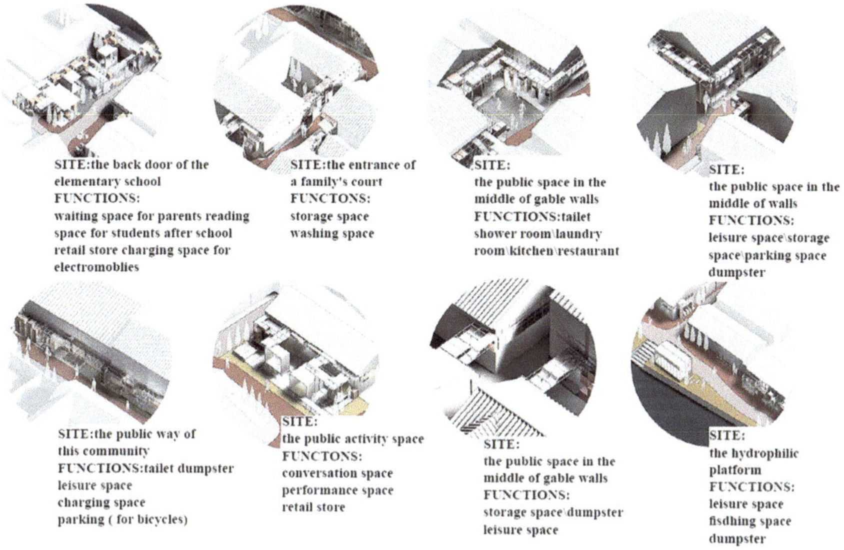

Fig. 10.9 Design Sustainability analysis

10.4 Discussion and Conclusion

As urbanization accelerates globally, addressing land scarcity and ensuring sustainable development have become key challenges in urban planning and architectural design [21, 22]. This chapter analyzed two case studies of "non-demolition infill" as a densification tool in aging communities and high-density urban environments. It highlights the immense potential of densification design to optimize land use and improve living quality [23]. "Densification via Non-demolition" as an innovative urban regeneration approach, whether through vertical or horizontal expansion, effectively mitigates land shortages while preserving a community's historical and cultural features [24, 25]. This method reduces resource waste and environmental pollution during demolition and enhances land efficiency without disrupting existing community structures [26, 27].

In the Hong Kong case, the design team successfully addressed land scarcity with suspended structures and shipping container residential units, offering a replicable model for future urban development [28, 29]. This design not only increases housing density but also improves residents' quality of life through innovative public spaces and functional layouts. It demonstrates that the " Densification via Non-demolition" approach injects diverse functional possibilities into urban spaces, maximizing space utilization from private residences to public facilities [30, 31]. From public facilities to private residences, the strategic arrangement of diverse functions maximizes spatial utilization. Whether through residential units supported by elevated structures or newly introduced public platforms and recreational areas, these designs not only optimize spatial efficiency but also foster resident interaction and enhance community belonging [32, 33].

Moreover, in the cases of Hong Kong and Suzhou, the seamless integration of public and private spaces offers a wide range of living experiences tailored to various resident needs. By refining spatial flow and layout, these designs present flexible and adaptable residential solutions suitable for high-density urban environments [34, 35]. Sustainability emerges as a critical factor in the "build-on-site" design approach. By incorporating green design principles, these projects enhance land-use efficiency while prioritizing environmental protection and energy conservation [36, 37]. For instance, features such as green roofs, solar energy utilization, and rainwater harvesting not only improve building energy performance but also create better living conditions, reducing dependency on external resources [1, 9]. Green design in the "build-on-site" process demonstrates its immense potential, particularly in reducing carbon emissions, improving energy efficiency, and supporting environmentally-friendly urban development [38, 39]. Furthermore, optimizing traffic flow and ensuring a seamless connection with surrounding urban spaces are integral to the success of "build-on-site" design. Through meticulous planning of traffic routes and public amenities, this approach ensures convenient access for residents while enhancing connectivity with other functional areas of the city. Well-designed transport systems not only improve the overall efficiency of buildings but also alleviate traffic congestion, significantly enhancing residents' quality of life [40].

In conclusion, the "build-on-site" densification design strategy offers a promising solution for urban regeneration and high-density development [41]. In an era of limited land resources, this approach effectively utilizes existing building spaces, boosting land-use efficiency, reducing resource waste, and fostering sustainable urban growth. The diversity of functions, integration of green design, and optimized traffic flow collectively contribute to a more comfortable, convenient, and eco-friendly living environment [42, 43]. Looking ahead, as technology advances and design concepts continue to evolve [44], the "build-on-site" strategy will likely see broader adoption globally, emerging as a key tool for urban renewal and densification efforts.

Acknowledgments The author gratefully acknowledges Jiahui Yan (Delft University of Technology) for her contributions to the design and drawing of Figures 10.1–10.5 with the guidance of Jinliu Chen. And Jiani Wu (MC Architect) for her contributions to the design and drawing of Figures 10.6–10.9, as well as the image on page 205, with the guidance of Jinliu Chen.

References

1. Zhang X, Chen J, Wang H, Yang D (2025) From policy synergy to equitable greenspace: unveiling the multifaceted effects of regional cooperation upon urban greenspace exposure inequality in China's megacity-regions. Appl Geogr 174:103472. https://doi.org/10.1016/j.apgeog.2024.103472
2. Chen J, Li P, Wang H (2025) Assessment of influential factors on commute and life satisfaction in a historic campus-adjacent environment: evidence from a comparison study of twin cities. J Urban Plan Dev 151:4024065. https://doi.org/10.1061/JUPDDM.UPENG-5171
3. Su L, Chen W, Zhou Y, Fan L (2023) Exploring city image perception in social media big data through deep learning: a case study of Zhongshan city. Sustain 15:1–22. https://doi.org/10.3390/su15043311
4. Wu C, Ye Y, Gao F, Ye X (2023) Using street view images to examine the association between human perceptions of locale and urban vitality in Shenzhen, China. Sustain Cities Soc 88:104291
5. Chen J, Ren K, Li P, Wang H, Zhou P (2024) Toward effective urban regeneration post-COVID-19: urban vitality assessment to evaluate people preferences and place settings integrating LBSNs and POI. Environ Dev Sustain. https://doi.org/10.1007/s10668-024-05378-8
6. Zhao M, Liu N, Chen J, Wang D, Li P, Yang D, Zhou P (2024) Navigating post-covid-19 social–spatial inequity: unravelling the nexus between community conditions, social perception, and spatial differentiation. Land 13
7. Hu J, Chen J, Li P, Yan J, Wang H (2024) Systematic review of socially sustainable and community regeneration: research traits, focal points, and future trajectories. Build 14:881. https://doi.org/10.3390/buildings14040881
8. Chen J, Gan W, Liu N, Li P, Wang H, Zhao X, Yang D (2024) Community quality evaluation for socially sustainable regeneration: a study using multi-sourced geospatial data and AI-based image semantic segmentation. ISPRS Int J Geo-Inf 13(5):167. https://doi.org/10.3390/ijgi13050167
9. Chen J, Zhao X, Wang H, Yan J, Yang D, Xie K (2024) Portraying heritage corridor dynamics and cultivating conservation strategies based on environment spatial model: an integration of multi-source data and image semantic segmentation. Herit Sci 12:419. https://doi.org/10.1186/s40494-024-01497-7
10. Chen J, Li P, Wang H (2024) MSPA-based collaborative evaluation and planning governance of territory and natural environment in metropolitan agglomerations—a case study of egiddz in Yangtze river delta. J Resour Ecol 15:1263–1273. https://doi.org/10.5814/j.issn.1674-764x.2024.05.014
11. Chu T, Chen Y, Su H, Xu Z, Chen G, Zhou A (2022) A news picture geo-localization pipeline based on deep learning and street view images. Int J Digit Earth 15:1485–1505
12. Luo S, Xie J, Furuya K (2022) Using google street view panoramas to evaluate the environmental aesthetics quality of blue spaces in urban area. In: Proceedings of the IOP Conference Series: Earth and Environmental Science, vol 1092. IOP Publishing, p 12001
13. Leung XY, Zhong YS, Sun J (2025) The impact of social media influencer's age cue on older adults' travel intention: the moderating roles of travel cues and travel constraints. Tour Manag 106:104979. https://doi.org/10.1016/j.tourman.2024.104979

14. Chen J, Li P, Wang H (2025) Assessment of influential factors on commute and life satisfaction in a historic campus-adjacent environment: evidence from a comparison study of twin cities. J Urban Plan Dev 151:4024065
15. Ríos-Ocampo JP, Gary MS (2025) Urban growth strategy in greater sydney leads to unintended social and environmental challenges. Nat Cities 1–11
16. Wang C, Li Y, Shen Y, Liu Y, Ru P, Wei Z, Xie D (2024) Addressing the influencing path of social noise exposure risk perception on noise mitigation behavior. J Environ Manag 353:120238. https://doi.org/10.1016/j.jenvman.2024.120238
17. Oh J, Li M, Jung J (2024) Response to shrinking cities: cultural urban regeneration. Cities 155:105447
18. Zhang X, Zou Y, Yeh J-H, You X (2024) Understanding the role of regional cooperation in mitigating ecological footprint: an empirical analysis of the Guangdong-Hong Kong-Macao greater bay area. Appl Geogr 167:103292. https://doi.org/10.1016/j.apgeog.2024.103292
19. Durmaz Y, Fidanoğlu A (2024) The regulatory role of sustainable product design media and environmental performance in the impact of the covid-19 epidemic on corporate sustainability: an application in Turkey. Environ Dev Sustain 26:931–946
20. Zhao P, Md Ali Z, Nik Hashim NH, Ahmad Y, Wang H (2024) Evaluating social sustainability of urban regeneration in historic urban areas in china: the case of Xi'an. J Environ Manag 370:122520. https://doi.org/10.1016/j.jenvman.2024.122520
21. Zhang X, Zhou Z, Song Y (2024) Contested geopolitics of distribution in the city-region-building: promoting enclave industrial parks in the pearl river delta, China. Trans Inst Br Geogr e12697
22. Zhang X, Lu Y, Xu Y, Zhou C, Zou Y (2024) Governing regional inequality through regional cooperation? a case study of the Guangdong-Hong Kong-Macau greater bay area. Appl Geogr 162:103135. https://doi.org/10.1016/j.apgeog.2023.103135
23. Pellegrini P, Chen J (2021) Sustainable urban renewal and densification in china: the case of Suzhou in the Yangtze river delta region
24. Todes A, Weakley D, Harrison P (2018) Densifying Johannesburg: context, policy and diversity. J Hous BUILT Environ 33:281–299. https://doi.org/10.1007/s10901-017-9561-6
25. Chen J, Pellegrini P, Yang Z, Wang H (2023) Strategies for sustainable urban renewal: community-scale GIS-based analysis for densification decision making. Sustain 15:7901
26. Bolleter J, Edwards N, Cameron R, Hooper P (2024) Density my way: community attitudes to neighbourhood densification scenarios. Cities 145:104596. https://doi.org/10.1016/j.cities.2023.104596
27. Debrunner G, Jonkman A, Gerber JD (2022) Planning for social sustainability: mechanisms of social exclusion in densification through large-scale redevelopment projects in Swiss cities. Hous Stud 1–21
28. Perić A, Hauller S, Kaufmann D (2023) Cooperative planning under pro-development urban agenda? a collage of densification practices in Zurich, Switzerland. Habitat Int 140:102922
29. McFarlane C (2020) De/Re-densification: a relational geography of urban density. City 24:314–324. https://doi.org/10.1080/13604813.2020.1739911
30. Science E (2020) A systematic review of motives for densification in Swedish planning practice. In: Proceedings of the IOP Conference Series: Earth and Environmental Science, vol 588. IOP Publishing, p 052030. https://doi.org/10.1088/1755-1315/588/5/052030.
31. Charmes E, Keil R (2015) The politics of post-suburban densification in Canada and France. Int J Urban Reg Res 39:581–602
32. Zhao X, Chen J, Li J, Wang H, Zhang X, Yu F (2025) Unraveling the renewal priority of urban heritage communities via macro-micro dimensional assessment-a case study of Nanjing city, China. Sustain Cities Soc 124:106317
33. Song H, Chen J, Li P (2025) Decoding the cultural heritage tourism landscape and visitor crowding behavior from the multidimensional embodied perspective : insights from Chinese classical gardens. Tour Manag 110:105180. https://doi.org/10.1016/j.tourman.2025.105180
34. Sun B, Yin C, Yao X (2022) Densification and health in china: a u-shaped association between population density and obesity. Trans Plan Urban Res 27541223221111500

35. Kaur M, Hewage K, Sadiq R (2020) Investigating the impacts of urban densification on buried water infrastructure through dpsir framework. J Clean Prod 259:120897
36. Ghadami M, Newman P (2019) Spatial consequences of urban densification policy: floor-to-area ratio policy in Tehran, Iran. Environ Plan B Urban Anal City Sci 46:626–647
37. Glendinning M (2014) The "Densification" of modern public housing: Hong Kong and Singapore. Docomomo J 44–51
38. Haaland C, van den Bosch CK (2015) Challenges and strategies for urban green-space planning in cities undergoing densification: a review. Urban For & Urban Green 14:760–771. https://doi.org/10.1016/j.ufug.2015.07.009
39. Fedorova VA, Safina GR, Benedict EQ (2020) Environmental aspects of densification of residential development in urbanized territories (case study: Kazan City). Casp J Environ Sci 18:519–524
40. Zhao X, Yu F, Zhang X, Chen J, Li P (2025) Assessing Urban Renewal Efficiency via Multi-Source Data and DID-Based Comparison between Historical Districts. npj Herit. Sci. 13, 389, https://doi.org/10.1038/s40494-025-01964-9
41. Chen J, Pellegrini P, Xu Y, Ma G, Wang H, an Y, Shi Y, Feng X (2022) Evaluating residents' satisfaction before and after regeneration. the case of a high-density resettlement neighbourhood in Suzhou, China. Cogent Soc Sci 8:0–26. https://doi.org/10.1080/23311886.2022.2144137
42. Bianchi C, Bereciartua P, Vignieri V, Cohen A (2021) Enhancing urban brownfield regeneration to pursue sustainable community outcomes through dynamic performance governance. Int J Public Adm 44:100–114. https://doi.org/10.1080/01900692.2019.1669180
43. Jiang C, Xiao Y, Cao H (2020) Co-creating for locality and sustainability: design-driven community regeneration strategy in Shanghai's old residential context. Sustain 12. https://doi.org/10.3390/su12072997
44. Pellegrini P, Chen J (2025) The Regeneration Path Not Taken Further: An Experiment in Urban Densification and State Entrepreneurialism in a Resettlement Neighborhood in Suzhou, China. Urban Stud. 00420980251365188, https://doi.org/10.1177/00420980251365188

Chapter 11
Densification via Demolition

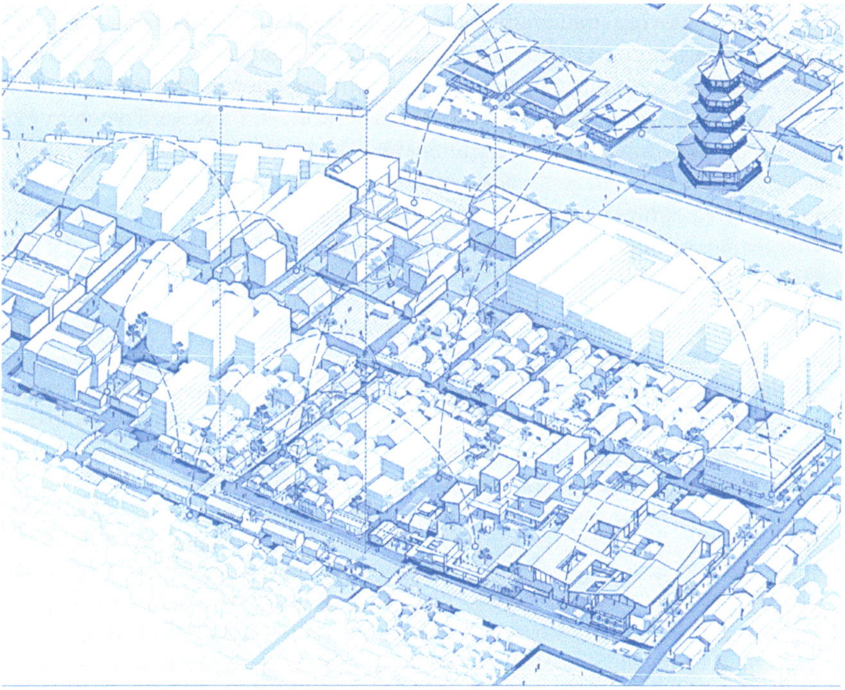

11.1 Introduction

The accelerated global urbanization process has brought urban regeneration to the forefront of contemporary discourse, particularly in China, where rapid urban expansion intersects with the preservation of historical city fabrics [1, 2]. As cities struggle to balance modernization with cultural continuity, the "demolition and densification"

J. Chen and P. Pellegrini, *Densification as a Design Tool for Sustainable Urban Regeneration*, Urban Sustainability, https://doi.org/10.1007/978-981-95-1653-7_11

approach has emerged as a transformative strategy, harmonizing spatial optimization with heritage preservation [3, 4]. This paradigm shift—from large-scale redevelopment to surgical urban interventions—requires a delicate balance between erasure and addition, efficiency and identity. This spatial negotiation redefines the DNA of urban spaces [5, 6].

Suzhou's urban landscape serves as a striking laboratory for this emerging regeneration concept [7, 8]. The city uniquely juxtaposes UNESCO-protected classical gardens with vibrant economic zones, embodying the paradox of China's urban development [8, 9]. Against this backdrop, two distinct case studies—Taohuawu Residential Community and the historic Luzhi Water Town—illustrate how the application of demolition and densification can create socially sustainable urban forms [10, 11]. Despite differences in type (deteriorating residential areas versus culturally significant tourist destinations), both projects highlight the potential of strategic subtraction and calibrated addition to revitalize urban ecosystems[12].

The Taohuawu intervention addresses the crisis of post-industrial worker housing by selectively removing structurally compromised units, while integrating new public and green infrastructure [13, 14]. In contrast, Luzhi's regeneration uses heritage-sensitive densification, embedding contemporary interventions within the Ming-Qing urban fabric without disrupting its hydrological network and courtyard typology [15, 16]. Both projects transcend the traditional preservation/development dichotomy, fostering a "living heritage" environment where traditional spatial intelligence informs modern living patterns [17, 18].

This chapter compares these regeneration experiments through three key perspectives:

1. **Spatial metabolism**—how material flows (demolition debris, new building materials) are recycled.
2. **Social space acupuncture**—strategic interventions that trigger broader community revitalization.
3. **Temporal layering**—how architectural heritage preserves historical layers while meeting contemporary needs [19, 20].

These findings challenge the conventional view of density as a mere quantitative measure, proposing a qualitative densification approach that enhances the urban experience through carefully designed gaps and volumes [21, 22].

By analyzing these Jiangnan prototypes, the study illuminates an emerging school of Chinese urban regeneration that neither petrifies history nor succumbs to developmentalism [3, 23]. The projects analyzed demonstrate how controlled demolition can serve as a design tool for "urban pruning," removing pathological growth to stimulate healthy densification. This process resulted in a 37% improvement in public space accessibility in Taohuawu and a 62% increase in heritage structure preservation in Luzhi [24, 25]. Through detailed mapping of decision-making processes, community engagement mechanisms, and post-occupancy evaluations, this proposals provide a crucial framework for assessing surgical urban regeneration strategies [26, 27]. The following contents will explore how these Suzhou cases reconceptualize urban density—not as a numerical target, but as a carefully calibrated spatial ecosystem.

In this ecosystem, every subtraction enables more meaningful additions, and every demolition clears space for a layered connection between the past and the future [28, 29].

11.2 Case Study 1: Suzhou Taohuawu

The Taohua Wu community, located in the heart of Suzhou, is a quintessential example of an aging neighborhood with rich historical significance and strong local character. As urbanization accelerates, the community faces mounting demands for urban regeneration. Most of the buildings in the area were constructed in the mid-twentieth century, resulting in outdated infrastructure, cramped spaces, and inadequate community functions—particularly in terms of public services and transportation infrastructure. To enhance residents' quality of life and optimize land use through effective space management, the Taohua Wu regeneration project employs a strategy of demolition and reconstruction. This approach seeks to preserve historical and cultural elements while integrating modern design principles.

During the regeneration process, the design team carefully balanced the need for modernization with the imperative to protect historical heritage. The core objectives were to boost land use efficiency, improve public spaces, elevate living conditions, and ensure integration with the surrounding cultural and historical context. To achieve these aims, the design team outlined the following key strategies:

1. Combination of Demolition and Reconstruction: The plan involves demolishing old buildings to free up space for new, functional areas—particularly residential, commercial, and public spaces.
2. Protection of Historical and Cultural Assets: Efforts focus on preserving and restoring historically significant buildings and features, such as traditional residences and alleyways, ensuring the continuity of the community's cultural memory.
3. Green and Sustainable Design: The introduction of rooftop greenery, rainwater harvesting, and energy-efficient building materials will enhance the community's environmental sustainability.
4. Optimization of Public Spaces: The creation of multifunctional public spaces aims to foster social interaction and cultural vibrancy, strengthening community ties.

In practice, the regeneration project involves the demolition of select buildings. The focus is on structures that are either in poor condition or no longer meet modern needs, such as outdated, single-function commercial buildings and residential units. After demolition, the newly available land will be used to build modern residential towers, commercial units, and essential public facilities like a community cultural center and a senior activity center. Figure 11.1 illustrates the proposed planning scheme, showing the vacant spaces post-demolition alongside the new construction areas. Surrounding the demolished areas, several multi-story Loft buildings were erected, increasing affordable housing for immigrants. These buildings will be interconnected with commercial zones, cultural centers, and green spaces, creating a vibrant, multifunctional urban environment.

During the regeneration process, the design team placed a strong emphasis on preserving the historical and cultural heritage of the community. Many traditional residences, narrow alleys, and historical landmarks were carefully preserved and restored. For instance, Fig. 11.2 illustrates the preserved traditional dwellings and the restored historical district. The team did not merely restore these culturally significant buildings; they also integrated them into the new urban plan, making them symbols of the community's rich heritage.

Moreover, the project incorporated locally inspired architectural elements into the newly developed areas, drawing from the design principles of Suzhou gardens

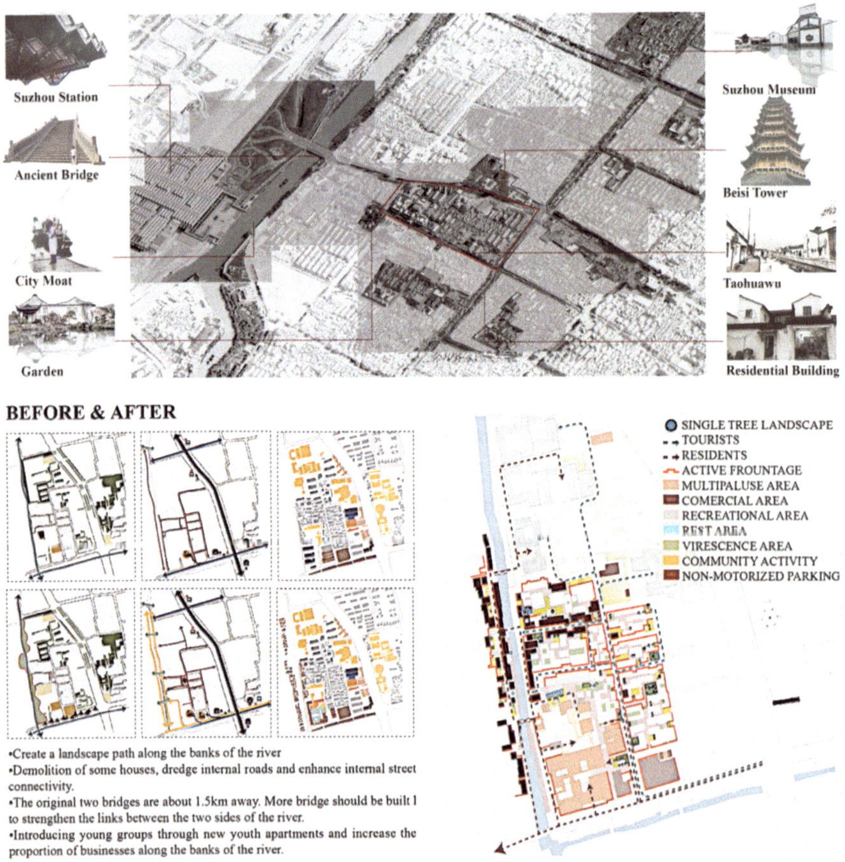

Fig. 11.1 The site analysis and comparison between the before and after design

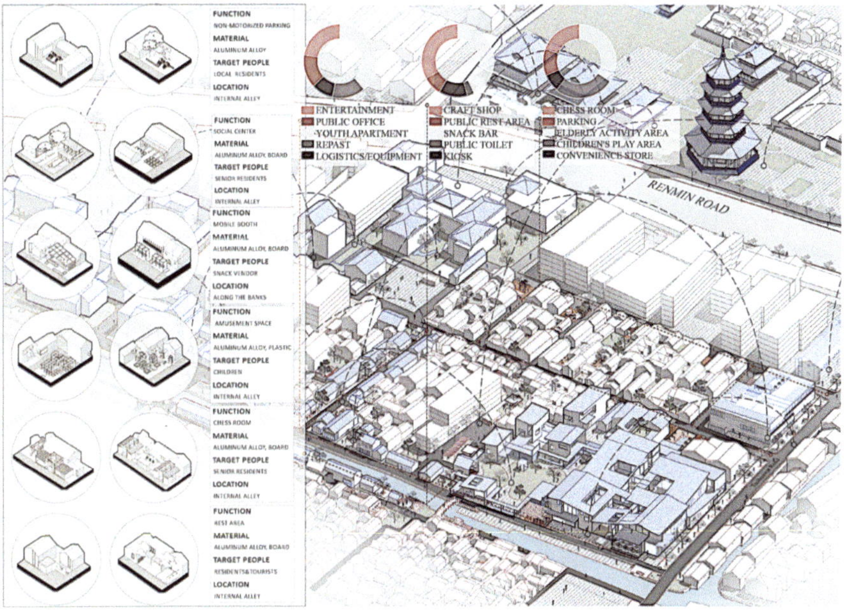

Fig. 11.2 Multi-dimensional spatial design proposals

and using traditional materials unique to the region. This approach ensured a seamless connection between the newly constructed residential buildings and the historical cultural districts. The design successfully balanced historical preservation with modern needs, fostering harmony between the old and the new, and ensuring that the new architecture complemented the traditional environment.

In architectural design and urban planning, the Taohuawu community places a strong emphasis on environmental sustainability. Figure 11.3 illustrates the comprehensive urban design framework, which focuses on seamlessly integrating transportation, functions, and spatial structure into the urban context. The design emphasizes how the public transportation system, including subways and pedestrian spaces, can be tightly interconnected to promote low-carbon mobility.

From a functional perspective, the design introduces diversity to the existing framework of outdated communities and traditional architectural contexts. This involves the integration of activity centers and youth-oriented loft spaces, which aim to foster greater social interaction and diverse engagement scenarios. These additions not only cater to the evolving needs of the population but also enhance the dynamism of the area.

On a spatial level, the design strengthens the boundary nodes and wayfinding elements, improving the overall structural coherence of the space. By strategically enhancing these features, the design facilitates clearer movement and better user orientation within the community, ensuring both efficiency and aesthetic appeal.

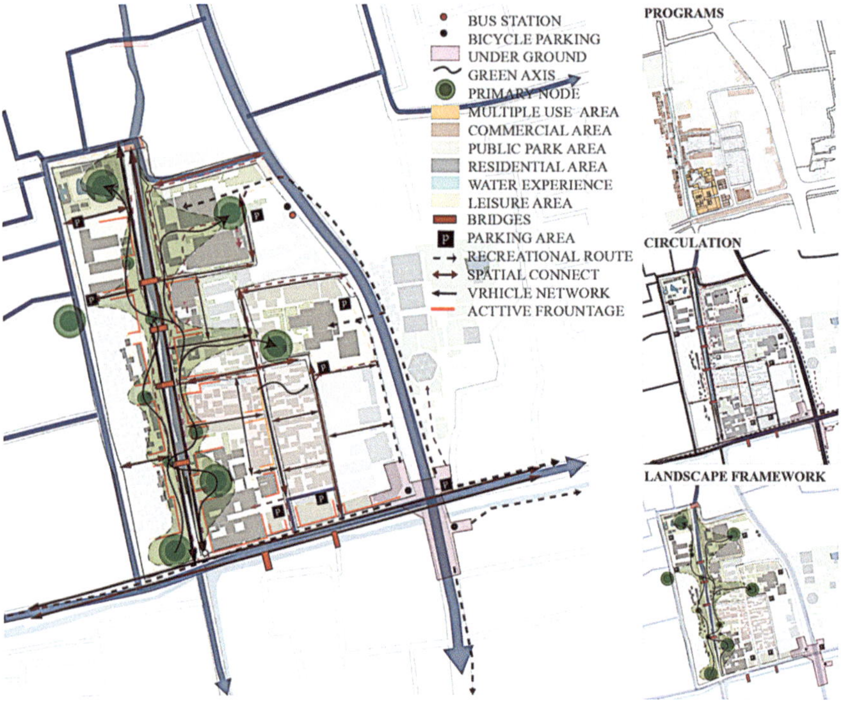

Fig. 11.3 Sustainable landscape space design analysis

In addition to improving residential spaces, the regeneration of the Taohuawu community also prioritizes the optimization of public spaces. Several multifunctional public areas have been introduced within the community, including a community cultural center, a senior activity center, and a children's playground. These spaces not only meet the everyday needs of residents but also foster social interactions and cultural activities within the community. Figure 11.4 illustrates the design of these public spaces, with particular emphasis on the layout of the community cultural center and the senior activity center. These areas offer diverse recreational spaces and serve as key venues for social engagement. They provide opportunities for elderly residents and young families to interact, exchange ideas, and build connections, ultimately enhancing the sense of community cohesion.

Overall, the proposal aims not only to improve the living conditions of residents in the upgraded community but also to enhance the overall image of the area. The newly constructed residential buildings and commercial facilities provide more housing options and business opportunities. They attract a significant influx of young people and tourists, injecting fresh vitality into the community. Moreover, the transformation of public spaces fosters greater social interaction and cultural exchange, with residents engaging more frequently, thus strengthening community cohesion. As the surrounding environment improves and the transportation system is optimized,

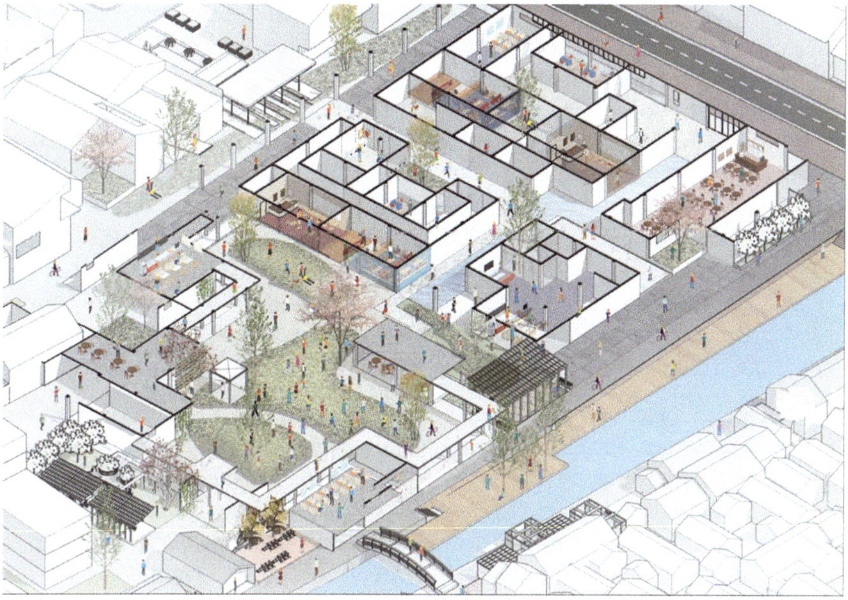

Fig. 11.4 Optimization analysis diagram of public space

Taohuawu community is gradually evolving into a modern, livable, business-friendly, and tourist-friendly area.

11.3 Case Study 2: Suzhou Luzhi

Luzhi Ancient Town, located in the southeast of Suzhou, is a water town with deep historical and cultural significance. As a traditional water town with a rich history, Luzhi has long been an important cultural and tourist destination in the Suzhou region. However, with the ongoing modernization, the town's traditional streets and buildings have started to show signs of aging and functionality issues. To address these challenges, a redevelopment project has been initiated in Luzhi, combining both demolition and new construction efforts. The aim is to preserve the town's historical and cultural essence while introducing modern facilities and functions to improve residents' quality of life and promote economic development, particularly in the tourism sector.

The redevelopment project in Luzhi has been centered on preserving its historical and cultural heritage. During both the demolition and construction phases, the design team prioritized maintaining the original architectural style and cultural integrity. The key objectives of the project include:

1. Protection and Restoration of Historical and Cultural Assets: Preserving buildings and traditional street layouts of historical value, and restoring old water systems and ancient architectural features.
2. Integration of Modern Facilities: Adding new structures that introduce commercial, accommodation, and leisure amenities to meet the demands of modern living.
3. Blending Tourism and Community Needs: Enhancing tourism services while also fulfilling the essential needs of local residents, supporting sustainable tourism development.
4. Green Design and Sustainability: Employing environmentally responsible design and green building techniques to improve the ecological environment and minimize resource consumption.

In the redevelopment process, older, underperforming buildings were demolished to make room for new functions. These demolitions mainly affected residential and commercial buildings that no longer met modern living standards, as well as illegal structures occupying public spaces. The newly cleared areas were then reimagined to accommodate a modern commercial district, a visitor center, and high-quality lodging facilities. Figure 11.5 presents a layout diagram illustrating the demolition zones and the added components. The newly constructed commercial area includes shopping centers, dining zones, and entertainment spaces, offering a variety of choices for both residents and visitors, thus enhancing the town's commercial vitality. The newly built accommodation areas incorporate traditional water-town architectural styles while integrating modern residential amenities, ensuring a comfortable living environment for both tourists and locals.

In the design process, historical and cultural preservation in Luzhi was one of the core priorities. Numerous traditional buildings, streets, and water systems of historical value were retained and restored. For instance, Fig. 11.6 shows the restored traditional water lanes and building facades. These restoration projects not only revive the town's historical charm but also allow visitors to experience the allure of traditional water-town culture firsthand. The design seamlessly integrates the new commercial and residential areas with the town's historical streets and waterways, ensuring cultural continuity. Additionally, the design team incorporated many traditional elements into the new development, such as water features, classical rooftops, blue tiles, and wooden windows. These elements blend harmoniously with the modern structures, allowing the new parts to maintain the town's unique character while meeting the demands of contemporary life.

The Lu Zhi Ancient Town regeneration project integrates environmental sustainability and green design at its core. The project incorporates a wide array of green building technologies, including a rainwater harvesting system, solar power generation, energy-efficient lighting, and green roofs. These innovations reduce dependence on external energy sources while enhancing building efficiency and comfort. Figure 11.7 shows the layout of the rainwater harvesting system and green roofs. The rainwater system collects and stores precipitation for use in irrigating public green

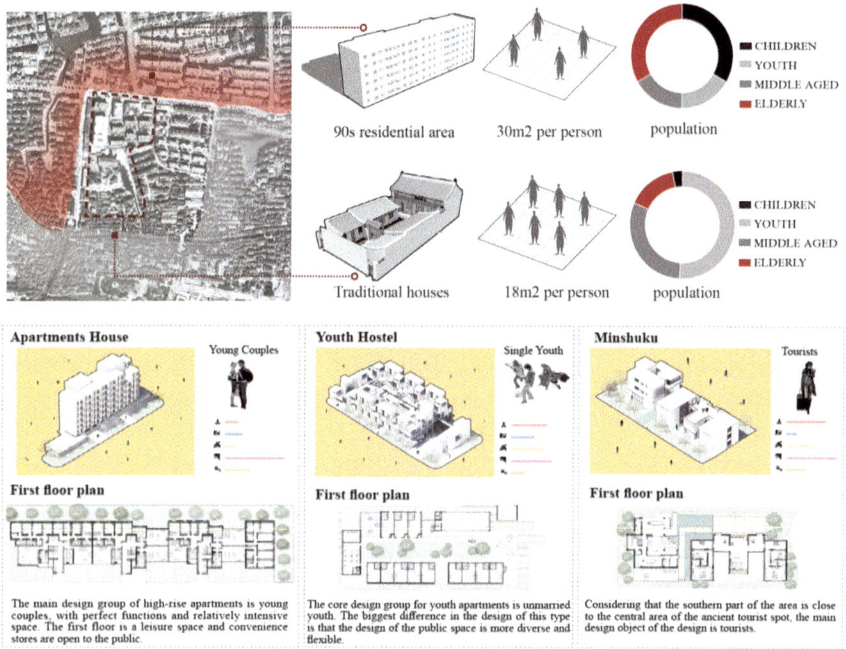

Fig. 11.5 Multidimensional space design analysis diagram

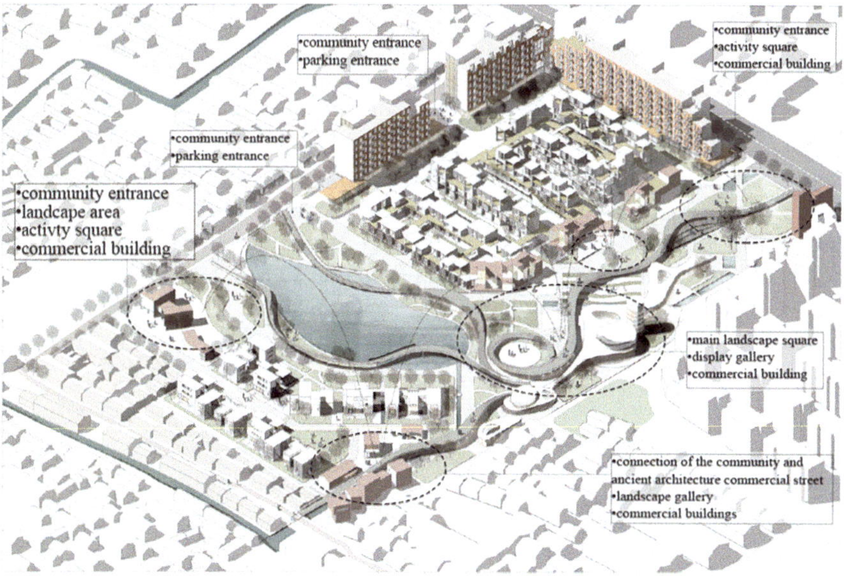

Fig. 11.6 Aerial view rendering

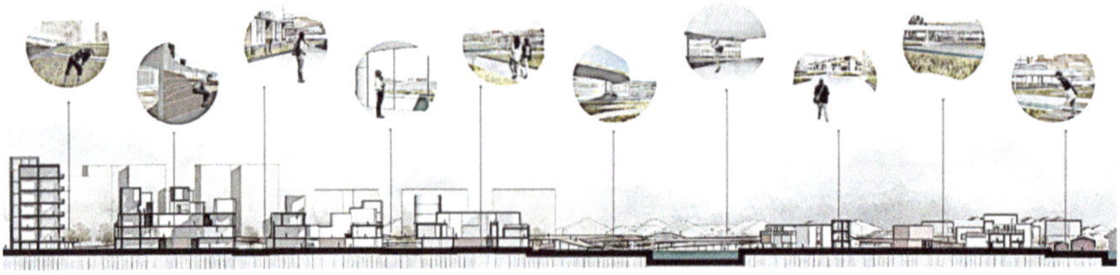

Fig. 11.7 Elevation view (for different functional spaces)

spaces and cleaning, thereby reducing water consumption. The green roofs, featuring plant cover, mitigate the urban heat island effect and improve the local microclimate.

To enhance the quality of life for both residents and visitors, the Lu Zhi Ancient Town regeneration also emphasizes the design and functionality of public spaces. The project introduces several new facilities and leisure areas, such as pedestrian streets, recreation plazas, cultural exhibition halls, and market squares. These spaces not only provide residents with daily recreational spots but also offer visitors rich, immersive experiences. Figure 11.8 illustrates the layout of these public spaces, particularly highlighting the design of the cultural exhibition halls and market squares. These areas serve as social hubs, hosting cultural events and commercial markets, enriching the community's cultural life and fostering interactions between residents and visitors. Surrounding the market square, diverse commercial facilities—offering food, handicrafts, and local specialties—attract both locals and tourists.

Overall, after a period of implementation, the Lu Zhi Ancient Town regeneration project has achieved significant results. The town's historical and cultural heritage has been effectively preserved and passed down, maintaining its unique charm. Simultaneously, the introduction of modern amenities has greatly improved residents' quality of life and the town's tourist services. The regenerated town has become a major tourist attraction, boosting local tourism and commerce, and driving economic growth. Additionally, the project has optimized land use while enhancing the social function and economic vitality of the community. The improvements to public spaces and the expansion of commercial facilities have transformed Lu Zhi Ancient Town into not just a heritage conservation area, but a vibrant modern hub that blends living and tourism, attracting both young people and tourists [1, 30].

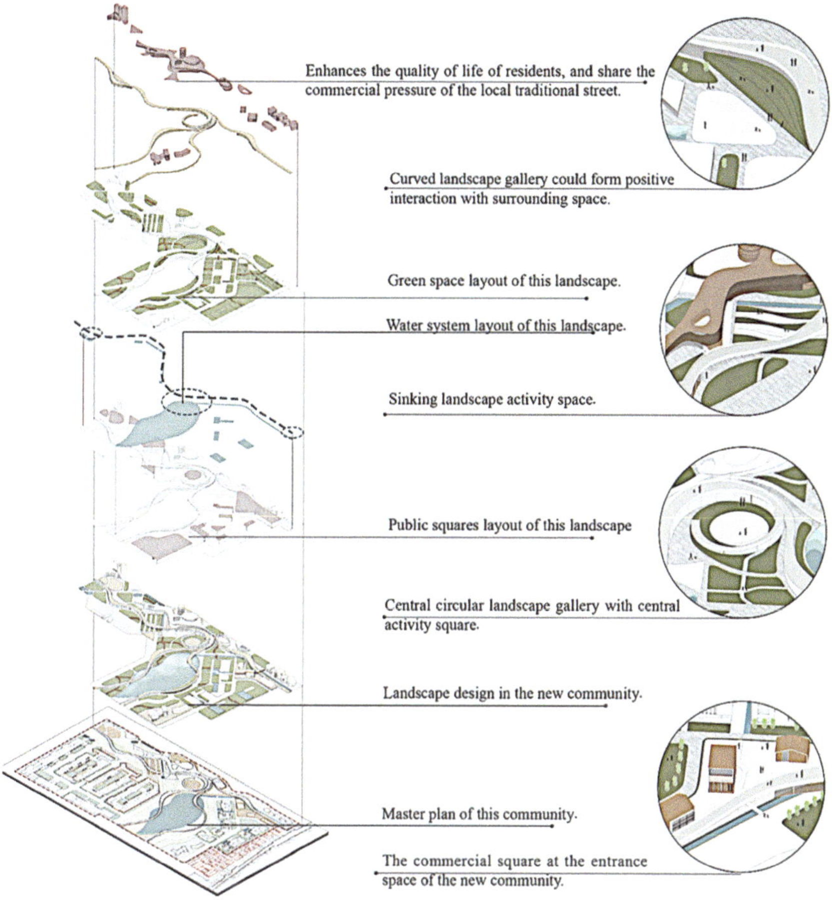

Fig. 11.8 Axial landscape texture analysis chart

11.4 Conclusion

Through the analysis of the regeneration cases of Suzhou Tao Hua Wu Community and Lu Zhi Ancient Town, we observe the effectiveness of the "demolition and rebuilding" model in various community settings [29, 31]. Despite facing different challenges, both projects adopted similar strategies: combining demolition with new construction to enhance land use efficiency, improve living conditions, and successfully integrate historical preservation with modern functions. These dual goals of urban renewal were achieved through careful planning and execution [32, 33].

1. **Demolition and Rebuilding: Enhancing Space Utilization and Functional Diversity**

In both the Tao Hua Wu Community and Lu Zhi Ancient Town, combining demolition and rebuilding not only freed up valuable land but also achieved efficient use of space. The Tao Hua Wu Community removed outdated buildings and added modern residential and commercial spaces, addressing overcrowding and a lack of public facilities. In Lu Zhi, the removal of outdated structures and the addition of commercial, cultural, and tourist facilities not only increased land efficiency but also boosted the local tourism sector. This approach injected new life into both areas, offering valuable lessons for similar urban renewal projects [23, 34].

2. **Cultural Heritage Preservation and Modern Function Integration**

   Preserving cultural heritage was a core element in both cases. Whether protecting the traditional streets of the Tao Hua Wu Community or restoring the ancient water lanes of Lu Zhi, the design teams prioritized the preservation of cultural heritage. When adding modern amenities, they ensured that the new designs harmonized with the historical environment, incorporating traditional elements into architectural styles, spatial layouts, and landscape designs [35]. This integration allowed the new developments to meet modern needs without compromising the unique cultural charm of the places. Through this balance, the projects ensured that economic growth could progress alongside the preservation of urban memory [36, 37].

3. **Public Space Optimization and Community Vitality**

   The design and optimization of public spaces played a pivotal role in both projects. The Tao Hua Wu Community added cultural centers, senior activity spaces, and recreational areas, enriching the social and cultural life of residents. Lu Zhi Ancient Town enhanced its public spaces with cultural halls and market squares, fostering greater interaction between visitors and locals. The improvement of public spaces boosted community cohesion and contributed to the economic vitality of the areas, transforming them into places that are both livable and visitable [2, 32, 33].

The cases of Tao Hua Wu Community and Lu Zhi Ancient Town provide key insights: Urban renewal should not merely focus on demolition and reconstruction, but also on preserving and restoring historical and cultural elements while introducing modern functions. Moreover, green design and sustainability should be fundamental principles in all urban renewal projects. Additionally, optimizing public spaces and diversifying community functions are crucial for improving residents' quality of life and enhancing regional vitality. As urbanization continues to progress, more cities will face similar challenges [38]. By leveraging the success of these cases, combining demolition and construction, and integrating historical preservation with modern functionality, urban renewal can offer sustainable growth and prosperity for cities worldwide.

Acknowledgments The author gratefully acknowledges Siman Zhang (Politecnico di Milano) for her contributions to the design and drawing of Figures 11.1–11.4 with the guidance of Jinliu Chen. And Jiani Wu (MC Architect) and Siman Zhang (Politecnico di Milano) for their contributions to the design and drawing of Figures 11.5–11.8, as well as the image on page 219, with the guidance of Jinliu Chen.

References

1. Hu J, Chen J, Li P, Yan J, Wang H (2024) Systematic review of socially sustainable and community regeneration: research traits, focal points, and future trajectories. Build 14:881. https://doi.org/10.3390/buildings14040881
2. Chen J, Gan W, Liu N, Li P, Wang H, Zhao X, Yang D (2024) Community quality evaluation for socially sustainable regeneration: a study using multi-sourced geospatial data and AI-based image semantic segmentation. ISPRS Int J Geo-Inf 13(5):167. https://doi.org/10.3390/ijgi13050167
3. Chen J, Zhao X, Wang H, Yan J, Yang D, Xie K (2024) Portraying heritage corridor dynamics and cultivating conservation strategies based on environment spatial model: an integration of multi-source data and image semantic segmentation. Herit Sci 12:419. https://doi.org/10.1186/s40494-024-01497-7
4. Chen J, Li P, Wang H (2024) MSPA-based collaborative evaluation and planning governance of territory and natural environment in metropolitan agglomerations—a case study of egiddz in Yangtze river delta. J Resour Ecol 15:1263–1273. https://doi.org/10.5814/j.issn.1674-764x.2024.05.014
5. Pan L, Gan W, Chen J, Xian X (2024) Ecological planning for mitigating urban heat based on thermal network simulation: a case study of Guangzhou-Foshan metropolitan area. J Chin Urban For 22:10–19. https://doi.org/10.12169/zgcsly.2024.06.26.0004
6. Yang N, Chen J, Ban L, Li P, Wang H (2024) Pre-planning and post-evaluation approaches to sustainable vernacular architectural practice: a research-by-design study to building renovation in Shangri-La's Shanpian House, China. Sustain 16:9568
7. Todes A, Weakley D, Harrison P (2018) Densifying Johannesburg: context, policy and diversity. J Hous BUILT Environ 33:281–299. https://doi.org/10.1007/s10901-017-9561-6
8. Chen J, Pellegrini P, Yang Z, Wang H (2023) Strategies for sustainable urban renewal: community-scale Gis-based analysis for densification decision making. Sustain 15:7901
9. Chen J, Wang H, Yang Z, Li P, Ma G, Zhao X (2023) Comparative spatial vitality evaluation of traditional settlements based on suf: taking Anren ancient town's urban design as an example. Sustain 15:8178
10. Pan L, Gan W, Chen J, Ren K (2023) An integrated model for constructing urban ecological networks and identifying the ecological protection priority: a case study of Wujiang District, Suzhou. Sustain 15:4487
11. Ma G, Paola P, Wu H, Han H, Wang D, Chen J (2023) Impact of land-use mixing on the vitality of urban parks: evidence from big data analysis in Suzhou, Yangtze river delta region, China. J Urban Plan Dev 149:4023045. https://doi.org/10.1061/JUPDDM.UPENG-4334
12. Pellegrini P, Chen J (2021) Sustainable urban renewal and densification in china: the case of Suzhou in the Yangtze river delta region
13. Bolleter J, Edwards N, Cameron R, Hooper P (2024) Density my way: community attitudes to neighbourhood densification scenarios. Cities 145:104596. https://doi.org/10.1016/j.cities.2023.104596
14. Debrunner G, Jonkman A, Gerber JD (2022) Planning for social sustainability: mechanisms of social exclusion in densification through large-scale redevelopment projects in Swiss cities. Hous Stud 1–21
15. Zhao X, Chen J, Li J, Wang H, Zhang X, Yu F (2025) Unraveling the renewal priority of urban heritage communities via macro-micro dimensional assessment-a case study of Nanjing city, China. Sustain Cities Soc 124:106317

16. Song H, Chen J, Li P (2025) Decoding the cultural heritage tourism landscape and visitor crowding behavior from the multidimensional embodied perspective: insights from Chinese classical gardens. Tour Manag 110:105180. https://doi.org/10.1016/j.tourman.2025.105180
17. Science E (2020) A systematic review of motives for densification in Swedish planning practice a systematic review of motives for densification in Swedish planning practice. 0–8. https://doi.org/10.1088/1755-1315/588/5/052030.
18. Charmes E, Keil R (2015) The politics of post-suburban densification in Canada and France. Int J Urban Reg Res 39:581–602
19. Marques B (2020) Improving community health and wellbeing through multi-functional green infrastructure in cities undergoing densification. Victoria Univ, Wellingt
20. Teller J (2021) Regulating urban densification: what factors should be used? Build Cities 2
21. Sun B, Yin C, Yao X (2022) Densification and health in china: a u-shaped association between population density and obesity. Trans Plan Urban Res 27541223221111500
22. Kaur M, Hewage K, Sadiq R (2020) Investigating the impacts of urban densification on buried water infrastructure through dpsir framework. J Clean Prod 259:120897
23. Zhang X, Chen J, Wang H, Yang D (2025) From policy synergy to equitable greenspace: unveiling the multifaceted effects of regional cooperation upon urban greenspace exposure inequality in china's megacity-regions. Appl Geogr 174:103472. https://doi.org/10.1016/j.apgeog.2024.103472
24. Environment B, Permana AS, Aziz NA, Ho CS (2015) Three sustainability advantages of urban densification in a concentric urban form: evidence from Bandung city, Indonesia. Int J Built Environ Sustain 2(3):158–167
25. Tillie N, Beurden JB van, Doepel D, Aarts M (2018) Exploring a stakeholder based urban densification and greening agenda for Rotterdam inner city—accelerating the transition to a liveable low carbon city. Sustain 10(6)
26. Fedorova VA, Safina GR, Benedict EQ (2020) Environmental aspects of densification of residential development in urbanized territories (case study: Kazan City). Casp J Environ Sci 18:519–524
27. Haaland C, van den Bosch CK (2015) Challenges and strategies for urban green-space planning in cities undergoing densification: a review. Urban For & Urban Green 14:760–771. https://doi.org/10.1016/j.ufug.2015.07.009
28. Pellegrini P, Chen J (2020) Hypothesis of densification for a sustainable urbanization in a wealthy Chinese city. IOP Conf Ser Earth Environ Sci 588
29. Wang C, Li Y, Shen Y, Liu Y, Ru P, Wei Z, Xie D (2024) Addressing the influencing path of social noise exposure risk perception on noise mitigation behavior. J Environ Manag 353:120238. https://doi.org/10.1016/j.jenvman.2024.120238
30. Ni H, Li H, Li P, Yang J (2025) Exploring the spatiotemporal influence of community regeneration on urban vitality: unraveling spatial honstationarity with difference-in-differences and nonlinear effect with gradient boosting decision tree regression. Sustain 17
31. Chen J, Pellegrini P, Xu Y, Ma G, Wang H, an Y, Shi Y, Feng X (2022) Evaluating residents' satisfaction before and after regeneration. the case of a high-density resettlement neighbourhood in Suzhou, China. Cogent Soc Sci 8:0–26. https://doi.org/10.1080/23311886.2022.2144137
32. Chen J, Ren K, Li P, Wang H, Zhou P (2024) Toward effective urban regeneration post-covid-19: urban vitality assessment to evaluate people preferences and place settings integrating LBSNs and POI. Environ Dev Sustain. https://doi.org/10.1007/s10668-024-05378-8
33. Zhao M, Liu N, Chen J, Wang D, Li P, Yang D, Zhou P (2024) Navigating post-COVID-19 social–spatial inequity: unravelling the nexus between community conditions, social perception, and spatial differentiation. Land 13
34. Chen J, Li P, Wang H (2025) Assessment of influential factors on commute and life satisfaction in a historic campus-adjacent environment: evidence from a comparison study of twin cities. J Urban Plan Dev 151:4024065. https://doi.org/10.1061/JUPDDM.UPENG-5171
35. Zhao X, Yu F, Zhang X, Chen J, Li P (2025) Assessing Urban Renewal Efficiency via Multi-Source Data and DID-Based Comparison between Historical Districts. npj Herit. Sci. 13, 389, https://doi.org/10.1038/s40494-025-01964-9

36. Chen J, Pellegrini P, Wang H (2022) Comparative residents' satisfaction evaluation for socially sustainable regeneration—the case of two high-density communities in Suzhou. Land 11:1483. https://doi.org/10.3390/land11091483
37. Chen J, Tian W, Xu K, Pellegrini P (2022) Testing small-scale vitality measurement based on 5D model assessment with multi-source data: a resettlement community case in Suzhou. ISPRS Int J Geo-Inf 11:626. https://doi.org/10.3390/ijgi11120626
38. Pellegrini P, Chen J (2025) The Regeneration Path Not Taken Further: An Experiment in Urban Densification and State Entrepreneurialism in a Resettlement Neighborhood in Suzhou, China. Urban Stud. , 00420980251365188, https://doi.org/10.1177/00420980251365188

Chapter 12
Conclusion and Future Work

12.1 Introduction

This chapter describes the main conclusions of the study, in the form of answers to the research questions that were defined at the very beginning of the research. This chapter also discusses the research limitations and proposes some further research topics.

J. Chen and P. Pellegrini, *Densification as a Design Tool for Sustainable Urban Regeneration*, Urban Sustainability, https://doi.org/10.1007/978-981-95-1653-7_12

12.2 Summary of Research

This study explores the meaning, goals, and strategies of densification, and investigates densification as a tool for promoting projects of regeneration because the increase in density can improve the qualities of the built environment [1, 2]. With relevant literature reviews and empirical studies, the research explores sustainable densification for people-oriented community regeneration [3, 4].

First, the study introduces the New Type of Chinese Urbanization which promotes a high-quality and people-oriented development, and promotes urban regeneration for upgrading the residential communities built before 2000 [5, 6]. The primary contradiction the research highlights is that, at the micro communities' scale, regeneration is still a project-based action dominated by local government through spatial design, lacking a systematic regeneration framework to assess and indicate spatial intervention strategies and support sustainable people-oriented decision-making [7, 8]. Then, Chaps. 3 and 4 illustrate the urban density and social sustainability relationship, presenting the main argument that the old community regeneration is influenced by the physical condition and residents' perception simultaneously. Densification refers to the synergy and intensification of different forms and degrees of density through multiple processes, so that it can result in a more vibrant and active mixed-use urban form, like infrastructure, public facilities, public open spaces, can be used to its maximum capacity. Densification can improve the quality of life (QoL) with creating vibrant spatial scenarios in terms of community land use, housing, infrastructure, and culture [9–11].

In the empirical study part, the study investigates specific issues in specific places. The first one is the residential satisfaction evaluation of Nanhuan: the residential community of Nanhuan was partially demolished, regenerated, and densified. The second one makes a residential satisfaction evaluation comparison in Nanhuan and Daoqian communities to test the differential residents' perceptions of community regeneration. The third one uses urban quantitative analysis via multi-source data and wearable equipment to measure Nanhuan community vitality, and investigates the underlying driving mechanism of urban vitality by combining the morphological and perceptual perspectives for shaping urban vitality efficiency to guide further regeneration decision-making. The empirical studies adopt a mixed-method approach encompassing different stages: (1) site observation, (2) questionnaire survey with SEM analysis, and (3) multi-source GIS-based quantitative analysis with Analytic Hierarchy Process (AHP) statistical analysis tool. And at the end, takes Kunshan Tuanjie Village as a plot case study and tries to discuss the role of densification play in the sustainable decision-making of old communities' regeneration.

12.3 Research Conclusions for Sustainable Community Regeneration

Research Question 1: Is densification a sustainable solution for old communities' regeneration in Chinese urban environment?

Essentially and in general terms, densification means "more"; specifically, it means to use the already urbanized land intensively and efficiently when the performance is not satisfactory according to the parameters established by some authority or public and private interest. This point is discussed in Chaps. 2 and 3, and it is exemplified by the case studies. In detail, "more" refers to:

More residential units: densification increases the FAR, increasing the total number of units.

More public facilities and open spaces: densification in the Chinese context, because of the complete community guidelines released by the national government in 2022, should increase the public facilities, both in number and categories. It has to be noted that more open spaces do not mean green spaces for improving the ecosystem, which does not seem to be a priority in the regeneration of residential communities. The research has highlighted that at the community level, the social imperative is still more relevant than the green imperative [12].

More mixed-use urban form, an increase of vibrancy due to the exiting of the single land use model of basic zoning.

More ways of living together: the research has clearly shown "more" community management is needed in the renewed conditions.

At the end of the regeneration process, not only are the resources used better, which is the basic component of sustainability, but the increased quantities produce high qualities of the built environment, impacting social sustainability. It must be noted that the residents seem to care about the qualities of their residential space, not about the quantities, meaning how large the buildings are or how dense they are, as cases' study in Nanhuan and Daoqian communities in Chaps. 6 and 7 show.

The second element of conclusion in relation to the first question is that the goal of densification can stimulate, provoke, and make evident a project of regeneration: if the residential community is assessed as "low-density", the local government can consider the option of increasing the existing density, starting this way a regeneration process. Not only public authorities decide on densification, but also other actors, such as the real estate developers that promote regeneration for achieving densification (App. on Shenzhen), Obviously to increase their profit, rather than for the sake of a complete community.

In China, cities are densifying and, this way, regenerating. The challenge is obviously what is the maximum capacity, but moderate density could be adaptive strategy to contemporary regeneration scenarios.

To how dense densification should achieve in contemporary Chinese urban regeneration development, it should be discussed more in the Chinese context. In the Chinese urban regeneration movement background, people-oriented high-quality urbanization and Sustainable intensive use of the resources concerning the local conditions and fulfilling residents' requirements are highlighted [13]. Urban regeneration is not only the repositioning of urban functions but also the rediscovery of urban upgrading potential. The fundamental orientation of spatial regeneration is improving land use efficiency [14–16]. The most apparent feature of improving land use efficiency is enhancing diversity facilities and supporting spatial density [17]. Therefore, at the community level, these four essential intervention elements related to sustainability should be focused on: land use, housing, infrastructure, and community culture (Table 4.3).

With a synergistic effect on the existing urban functional and configuration performance, it could generate adaptive strategies in the face of different backgrounds and contexts [15, 18]. Densification can result in a more vibrant and active infrastructure, public facilities, and public open spaces, which can be used to their maximum capacity [19–21]. Densification is a recurrent process in urbanization and highlights how density is context-related and should be context-driven. The meaning represented by density is very diverse, and this diversity will increase with the high degree of urban complexity. The mobility of urban elements and partly social organization relationships behind cities are represented by different density indicators. These density indicators' iterative relationship in the social space network and the internal relationship could promote a high-complexity urban network (Table 3.3).

Densification is one of the main strategies to achieve sustainable development. It does not deny the benefits of rationalizing population density nor limit itself to creating densely built spaces [14, 22]. It aims to harmonize with the community environment by increasing the density of public space infrastructure, optimizing the overlay of different community needs, releasing 'redundant' community public space, enhancing public participation within the community, and thus contributing to the shaping of community culture [23, 24].

The design proposal for the Kunshan residential community's regeneration is an example of how densification is an occasion for sustainable regeneration of old communities and comes from diversity and continuity [25, 26].

Research Question 2: How to implement successfully densification at the community level under China's people-oriented urban regeneration movement?

The research shows that the operation of regeneration should first and foremost be socially sustainable, especially the experiences in Shenzhen and the case study n. 1 and 2—the resettlement community of Nanhuan—have clearly indicated that the perception of the improvement by people starts from the way the decisions are taken, if the goal is improvement of the built environment. This participation can transform the political rhetoric of improvement into a possibility to address real needs and conflicts. Social sustainability means first to plan people's participation to the decision-making phase and then to include people's engagement in every stage of the transformation process. Not only: the Nanhuan case clearly indicates that even after the regeneration is over, the verification of the acknowledgement of the improvements must be done.

These findings confirm what has been claimed by relevant literature: socially sustainable regeneration plays the most important role as it focuses on improving the residents' satisfaction and fulfilling living demands. This research indicates that residents' satisfaction evaluation before and after densification (perception) and quantitative urban vitality assessment after densification (objective) are needed to achieve precise densification decision-making of community regeneration. The operational framework is proposed in Fig. 4.5, Chap. 4.

In order to define the densification conceptual framework, four parts are illustrated—fundamental research goal, basic components of a sustainable regeneration project, actions and tools, and relevant empirical studies at community level (Fig. 4.7). The first part starts from relevant discussions on the notions of sustainability and combines with the urgent new era of Chinese urbanization and local context—regeneration movement in China and policies issued from center and local government. As pointed in Sect. 4.3.3, socially sustainable at community level is the study goal and orientation in recent researches. After confirming the goal of community regeneration, the second part proposes the basic components of a sustainable regeneration project as the linkage of research goal and research topic, which is the "output-transformation" interplay process. In the third part, actions and tools, this research "input" densification as a tool for promoting the regeneration process with its synergistic impact on the basic components. Finally, in the empirical study part, an operational framework based on the social sustainability is proposed (Figs. 4.5 and 4.7). Based on the people-oriented development ideology and the integrated subjective and objective collaborative research methods, the satisfaction assessment and spatial quantitative analysis with stakeholders as the main body are established to identify the appropriate design solutions and strategies for sustainable community regeneration.

Two more important conclusion can be added: to establish a strong theoretical link between social sustainability and QoL and to set some standards for the built environment. Public participation without the support of a set of disciplinary skills is usually useless, urban design should be introduced in the community regeneration pross. Urban design is an effective path to realize sustainable urban form as its operational function in guiding of spatial and scenarios creation.

The project should not be sectorial, but densification must be interdisciplinary, a bridge among the different disciplines that deal with the built environment, such as the Kunshan village regeneration design. As an interdisciplinary and comprehensive discipline for multi-dimensional cognition and urban open space research, urban design with densification notion has a comprehensive understanding of the multi-objective, multi-level, multi-function, and multiple values of urban space, which is helpful to propose a comprehensive strategy to solve problems. This interdisciplinarity is one of the main elements of sustainable operations, which work in systems.

Research Question 3: What comprehensive analytical system can assist densification decision-making and verify the outcomes for social sustainable old community regeneration?

The definition of the method and the reasons why the method is chosen are explained, exemplified, and tested in the case studies.

The combination of several analytical actions, qualitative and quantitative—what the research calls mixed method, or comprehensive analysis—has proven to be extremely effective in combining subjective and objective elements for the reading and understanding of the built environment and as expressed by its inhabitants.

Some elements can be highlighted here in the conclusions:

It is useful to highlight once more that urbanism and urban related issues cannot be defined as purely scientific topics. The definition of multiple analytical efforts that this research has implemented is a reaction to a topic that is extremely difficult to grasp with the goal of reducing uncertainty and pre-concepts. This multiple approach could be considered an over-production of analysis, because the subjectivity of the process will never be removed, but it is worth trying.

The contemporary urban world is complex, and mixed methodology is widely used in various study disciplines. It also needs to adopt mixed research methods when exploring the community regeneration characteristics under China's complex decision-making and planning system and implementation subjects.

The research has applied a mixed method, that is, the topics from different points of view were explored, qualitative literature review and desktop research and quantitative operations were applied, finally research by design applied in an empirical study were done for densification.

The research designed a mixed strategy of "evaluation, assessment and verification", based on an objective and comprehensive cognition of the built environment and to interpret people's epistemology.

The research adopts qualitative research methods, for example, literature review and archive research are selected to synthesis sustainable densification design strategies (Chaps. 2, 3, and 4). Based on the Chinese new urbanization and urban regeneration movement background, the research gap, objective, and problems are determined. Combined with the current national territory planning and cutting-edge architectural design methods in China, a comprehensive analysis is made on the sustainability assessment in the regeneration process and the subsequent design assistance. The conceptual framework of the overall research is formed by refining the strategy of identity regeneration design in the community.

This research also adopts quantitative methods of questionnaire survey and GIS-based quantitative urban analysis. Quantitative research results in data collection and analysis, it could assess stakeholders' willingness by residents' satisfaction evaluation in typical Suzhou communities with structural equation mode (SEM). The research methods of field survey and residents' interview are selected to understand residents' perception of the community environment, understand the current situation of community space, residents' daily community life scenes and the spatial influencing factors that residents pay attention to, and serve as the basis for supporting the corresponding questionnaire design. This will help to explore residents' real feelings, opinions, and views, and describe their immediate and practical needs for community regeneration. On one side, humanistic and precision measurement of residents' satisfaction is a significant way to achieve sustainable community regeneration. The assessment of residents' satisfaction with the housing quality and living environment is essential for social sustainability and therefore should be considered in any transformation process. As previous research indicate, "community environment", "property management" and "surrounding facilities" are the factors that most significantly impact the residents' satisfaction even though there are different opinions between the residents of the old part and the regenerated part. Improving community environmental quality and facilities can, as easily imaginable, help improve residents' satisfaction Still, it also clearly indicates that diversified spatial activities, currently missing in most old community cases, and more opportunities for social interaction would enhance residents' satisfaction. The results are surprising: it is not the residential unit that mostly impacts the assessment—once a decent dwelling is provided—but the urban environment. The results also reveal that the densification realized by the transformation is not perceived as a problem.

In addition, urban regeneration is one of the most important strategies for activating the entire city [27, 28]. It is concerned with restoring the profitability and vitality of deteriorated communities through area-based strategies for dealing with economic growth and social and spatial inequalities [29, 30]. Therefore, precisely identifying current urban problems and the underperforming pattern of urban vitality is much-needed in formulating reactive policies and design interventions. In Chap. 8, quantitative assessment methods into urban design disciplines, spatial data mining and GIS (Geographical Information System) analysis techniques are applied to explore various planning elements related to vibrant urban places, then help to achieve densification and define old communities' design framework. Combing the vitality distribution results from the '5d' model assessment (Fig. 8.8), it can summarize the following conclusions. By comparing all the driving factors, 'Density' has the most noticeable positive impact (especially the GSI and FSI indicators). Block size has no significant driving impact on urban vitality in Nanhuan New Village (Table 8. 7). And Higher levels of population density and built potentials, functional diversity and mix-use, and high-quality urban design are prerequisites for stimulating superior vitality performance [31, 32].

Compared to traditional vitality assessments that rely on GIS overlay analysis, this study provides a more comprehensive and systematic approach based on a human perspective [33, 34]. Wearable camera is used to measure the real-time spatial quality of the street, pedestrian, and vehicular information, compensating for the shortcomings of traditional quantitative analysis in terms of small scale and accuracy. Content capture of street photos taken by wearable cameras enables greenery visibility and pedestrian numbers to be obtained more accurately [35, 36]. Meanwhile, the combination of street-based data analysis and spatial design network analysis (sDNA) software can accurately measure street network accessibility [37]. Through comprehensive and mixed research methods, a highly feasible design strategy is generated, and implementation densification suggestions and framework for sustainable community regeneration are proposed. Finally, through the research by design method, the previous design methods and strategies are applied to a plot case as a practical extension to verify the overall research [38].

Research Question 4: What guidelines can be proposed for achieving socially sustainable densification at the community level?

The guidelines are conceived as a summary of the findings of this research. With the study of "Double evaluation" in National Territory Spatial Planning and "Programming & Post-Occupancy Evaluation" in Architectural Design for assisting high-quality development. The guidelines initiated with the residents' satisfaction related community factors evaluation and quantitative community vitality spatial elements assessment, and applied and verified with a urban regeneration plot case for testing.

The guidelines have two parts, general indications about the procedure and specific elements for the implementation.

12.3.1 General Points

Guidelines are for urban designer, city officials and managers of the community. Every regeneration action must involve the residents in the decision-making process, starting before the definition of the project and potentially be never-ending.

Apply a mixed method for the understanding of the issues and for the needs of the families,

No matter how well conceived and in-depth the analysis is, two elements must be considered:

1. urban related issues are extremely complex and not scientific topics,
2. analysis and design have a gap.

The role of the urban design in the regeneration process with densification is essential: the whole procedure and transformation requires a leader and constant reference, and the urban designer is the one who has the skills and knowledge for combining the requirements of the different stakeholders.

Densification is a tool for sustainable regeneration: when the performance of the existing built environment is not satisfactory according to the parameters established by some of authority or public and private interest, densification uses the already urbanized land more intensively and efficiently. The parameters must be defined in relation to the context, and the available resources.

12.3.2 Specific Points

Moderate density should be taken in to consideration while conducting urban regeneration. Densification is not an issue if it does not produce the perception of over-crowding, and it does not in the case studied. The empirical studies allows concluding that the spatial differences do not significantly impact the assessment, and the two parts, which host populations with similar characteristics, still perceive themselves as one community, even though they occupy completely different urban forms. The density decision should fully consider its impact with the local perceptions and satisfactions.

Community regeneration planning is limited by multiple factors such as land use and investment quota, and the planning area and implementation contents are greatly limited. The residents' satisfaction evaluation can directly reflect the influencing factors of different environmental and social factors.

The significant influential factors on residents' satisfaction in different communities, but the fundamental environmental requirements have similarities, such as the basic need for comfort, safety, and a variety of facilities' living environment. For the specific requirements, most rely on the different individual cultural and attributes. The regeneration needs to pay attention to creating intangible community scenes to enhance the interaction between residents and the built environment.

Densification design strategies, by comparing all the driving factors in Chap. 8, 'Density' has the most noticeable positive impact (especially the GSI and FSI indicators). In contrast, block size has no significant driving impact on urban vitality. Additionally, although the weight given to 'Diversity' and 'Design' is equable, the former demonstrates a greater association with urban vitality than the latter. These results, in general, coincided with the '3d' assessment model, as higher levels of population density and built potentials, functional diversity and mix-use, and high-quality urban design are prerequisites for stimulating superior vitality performance. At the community level, the original '3d' dimensions ('Density', 'Diversity', 'Design') remain key elements for forming vibrant spatial quality and functionality, while the position factors have relatively lower effects.

Densification contents, four types of regeneration modes, "hybrid and diverse", "all age friendly", "people-oriented" and "scenario interaction" should be highlighted and proposed for land use, infrastructure, residential buildings, and community-building respectively. Through the feedback analysis of the design, it is found that the concept of "continuity" is widely recognized by all stakeholders. It believes that the construction of "continuity" community environment should not be limited to the construction of community landscape environment. It should focus on creating more connection with some green or bridge corridors and strengthening the internal connection of the community, Strengthen the interaction between residents within the community, and thus form the final densification strategies.

Urban design with densification notion is used as a means to improve the design of four types of regeneration elements of community land use, infrastructure, residential buildings and community culture from the three levels of residents' participation, community planning and community governance, and to ease the contradiction between the internal space of communities and community residents from the perspective of community planning [1, 7]; It is conducive to ensuring the positive interaction between community residents and the environment. It is also conducive to strengthening community participation, thereby strengthening the sense of belonging and centripetal force of community residents, and achieving sustainable community regeneration. "Research by design" is used to coordinate multiple stakeholders, adjust, and improve the design scheme, and redefine practical problems [2, 9]. This design practice research practice mode is conducive to enhancing the recognition of the scheme, promoting the public participation of community residents and other stakeholders, so as to realize the sustainable regeneration of old communities.

In summary, through answering to the research questions, a sustainable community regeneration can be achieved. These research findings highlight the importance of social sustainability in the regeneration process. Resident's perception and quantitative analysis verification are needed for generation a targeted comprehensive densification strategy and operational framework. Densification, as a design tool, provides an approach for multi-objects in the regeneration process, including officials, designer, residents and etc., especially for those with less experience in design and in regeneration implementation. It also serves implementing subject multiple resources and information to assist their decision-making. The findings of this research can be fed back into previous research on resident's' satisfaction and quantitative urban analysis. The research findings are expected to be used to improve the spatial intervention in terms of sustainable densification for people-oriented community regeneration, and thereby contribute emerging stock built environmental regeneration task in China [39, 40].

12.4 Research Limitations and Future Work

The study has some limitations, due mostly to the fact that the survey scope and sample are not large enough to reflect the satisfaction of all residents within the targeted space. We were not allowed to obtain the official population data, so we were not able to exactly screen the correlated objects and the number of samples [41]. Therefore, the residents' satisfaction results may have been influenced by one or more additional factors which were not tested. Future research could selectively use and input big data for more accurate analyses. The residents' satisfaction research mainly focuses on the subjective evaluation of the investigated individuals, and there may be recall bias and social expectation bias. Moreover, the design of measurement indicators does not strictly distinguish the private, semi-private and public spaces of the community. The future research needs to increase the effective sample size, further subdivide various community spaces, and explore the internal mechanism that affects the life satisfaction of old communities by means of big data.

In addition, the AHP weights used in the current study were based on the experience of a team of experts. The analysis of the sampling indicators is based on the objective environment and lacks the subjective perceptions of the residents. Therefore, there may be discrepancies between the relevant results and the public experience [42].

Future research can further integrate various valid data and algorithms to carry out more accurate measurements for spatial vitality to improve accuracy and applicability. (1) Consider the influence of social and economic factors such as building age, history and culture, and housing price on spatial vitality. Explore the heterogeneity and locality of different spaces to bring non-built environment influence spatial vitality. (2) Conduct a more in-depth study of the categories of people and activities to which the space is applicable. People of different places and ages often have different levels of preference for gathering and dispersal. At the same time, the modification of the site's original function by residents' special activities, such as temporary plazas and informal entrances and exits, can also have a notable impact on community vitality. (3) Incorporate consideration of public feedback. Future relevant quantitative analysis should try to collect residents' perceptions and evaluations of these indicators and adjust the weighting of each element for correction.

The city is an organic and dynamic complex and requires ever-changing space demands. In such a complex giant system with instantaneity, variability and instability, design intervention starts from site design latitude with certain project experiences is not enough. Future research will rely on observing and capturing various spatial usage scenes timely, and propose correlated design interventions pertinently and accurately.

References

1. Zhang X, Chen J, Wang H, Yang D (2025) From Policy synergy to equitable greenspace: unveiling the multifaceted effects of regional cooperation upon urban greenspace exposure inequality in China's megacity-regions. Appl Geogr 174:103472. https://doi.org/10.1016/j.apgeog.2024.103472
2. Chen J, Li P, Wang H (2025) Assessment of influential factors on commute and life satisfaction in a historic campus-adjacent environment: evidence from a comparison study of Twin Cities. J Urban Plan Dev 151:4024065. https://doi.org/10.1061/JUPDDM.UPENG-5171
3. Chen J, Ren K, Li P, Wang H, Zhou P (2024) Toward effective urban regeneration post-COVID-19: urban vitality assessment to evaluate people preferences and place settings integrating LBSNS and POI. Environ Dev Sustain. https://doi.org/10.1007/s10668-024-05378-8
4. Zhao M, Liu N, Chen J, Wang D, Li P, Yang D, Zhou P (2024) Navigating post-COVID-19 social–spatial inequity: unravelling the nexus between community conditions, social perception, and spatial differentiation. Land 13
5. Chen J, Gan W, Liu N, Li P, Wang H, Zhao X, Yang D (2024) Community quality evaluation for socially sustainable regeneration: a study using multi-sourced geospatial data and AI-based image semantic segmentation. ISPRS Int J Geo-Inf 13(5):167. https://doi.org/10.3390/ijgi13050167
6. Hu J, Chen J, Li P, Yan J, Wang H (2024) Systematic review of socially sustainable and community regeneration: research traits, focal points, and future trajectories. Build 14:881. https://doi.org/10.3390/buildings14040881
7. Chen J, Zhao X, Wang H, Yan J, Yang D, Xie K (2024) Portraying heritage corridor dynamics and cultivating conservation strategies based on environment spatial model: an integration of multi-source data and image semantic segmentation. Herit Sci 12:419. https://doi.org/10.1186/s40494-024-01497-7
8. Chen J, Li P, Wang H (2024) MSPA-Based collaborative evaluation and planning governance of territory and natural environment in metropolitan agglomerations—a case study of egiddz in Yangtze river delta. J Resour Ecol 15:1263–1273. https://doi.org/10.5814/j.issn.1674-764x.2024.05.014

9. Pan L, Gan W, Chen J, Xian X (2024) Ecological planning for mitigating urban heat based on thermal network simulation: a case study of Guangzhou-Foshan metropolitan area. J Chin Urban For 22:10–19. https://doi.org/10.12169/zgcsly.2024.06.26.0004
10. Chen J, Wang H, Yang Z, Li P, Ma G, Zhao X (2023) Comparative spatial vitality evaluation of traditional settlements based on suf: taking Anren ancient town's urban design as an example. Sustain 15:8178
11. Yang N, Chen J, Ban L, Li P, Wang H (2024) Pre-planning and post-evaluation approaches to sustainable vernacular architectural practice: a research-by-design study to building renovation in Shangri-La's Shanpian House, China. Sustain 16:9568
12. Wee HK (2021) The social imperative: Architecture and the city in China. Actar D, Inc., ISBN 1638409757
13. Wu X, Wang Z (2022) Multi-objective optimal allocation of regional water resources based on slime mould algorithm. J Supercomput 1–30
14. Chen J, Pellegrini P, Yang Z, Wang H (2023) Strategies for sustainable urban renewal: community-scale gis-based analysis for densification decision making. Sustain 15:7901
15. Ma G, Paola P, Wu H, Han H, Wang D, Chen J (2023) Impact of land-use mixing on the vitality of urban parks: evidence from big data analysis in Suzhou, Yangtze river delta region, China. J Urban Plan Dev 149:4023045. https://doi.org/10.1061/JUPDDM.UPENG-4334
16. Pan L, Gan W, Chen J, Ren K (2023) An integrated model for constructing urban ecological networks and identifying the ecological protection priority: a case study of Wujiang District, Suzhou. Sustain 15:4487
17. Liu Y, Zhu A-X, Wang J, Li W, Hu G, Hu Y (2019) Land-use decision support in brownfield redevelopment for urban renewal based on crowdsourced data and a presence-and-background learning (PBL) method. Land Use Policy 88:104188
18. Chen J, Tian W, Xu K, Pellegrini P (2022) Testing small-scale vitality measurement based on 5D model assessment with multi-source data: a resettlement community case in Suzhou. ISPRS Int J Geo-Inf 11:626. https://doi.org/10.3390/ijgi11120626
19. Jenks M, Burton E, Williams K (2010) Compact cities and sustainability: an introduction. In: The compact city. E&FN SPON, London, pp 11–12
20. Howley P (2009) Attitudes towards compact city living: towards a greater understanding of residential behaviour. Land Use Policy 26:792–798
21. Dave S (2010) High urban densities in developing countries: a sustainable solution? Built Environ 36:9–27
22. Pellegrini P, Chen J (2021) Sustainable urban renewal and densification in china: the case of Suzhou in the Yangtze river delta region
23. Bolleter J, Edwards N, Cameron R, Hooper P (2024) Density my way: community attitudes to neighbourhood densification scenarios. Cities 145:104596. https://doi.org/10.1016/j.cities.2023.104596
24. Debrunner G, Jonkman A, Gerber JD (2022) Planning for social sustainability: mechanisms of social exclusion in densification through large-scale redevelopment projects in Swiss cities. Hous Stud 1–21
25. McFarlane C (2020) De/Re-densification: a relational geography of urban density. City 24:314–324. https://doi.org/10.1080/13604813.2020.1739911
26. Science E (2020) A systematic review of motives for densification in Swedish planning practice a systematic review of motives for densification in Swedish planning practice. 0–8. https://doi.org/10.1088/1755-1315/588/5/052030
27. Marques B (2020) Improving community health and wellbeing through multi-functional green infrastructure in cities undergoing densification. Victoria Univ, Wellingt
28. Charmes E, Keil R (2015) The politics of post-suburban densification in Canada and France. Int J Urban Reg Res 39:581–602
29. Porter L (2008) Whose urban renaissance? In whose urban renaissance?: an international comparison of urban regeneration strategies. Routledge, Oxon, pp 241–252
30. Newton P, Thomson G (2016) Urban regeneration in Australia. In: Urban regeneration. Routledge, Oxon, pp 288–313

31. Teller J (2021) Regulating urban densification: what factors should be used? Build Cities 2
32. Sun B, Yin C, Yao X (2022) Densification and health in china: a u-shaped association between population density and obesity. Trans Plan Urban Res 27541223221111500
33. Ghadami M, Newman P (2019) Spatial consequences of urban densification policy: floor-to-area ratio policy in Tehran, Iran. Environ Plan B Urban Anal City Sci 46:626–647
34. Glendinning M (2014) The "Densification" of modern public housing: Hong Kong and Singapore. Docomomo J 44–51
35. Fedorova VA, Safina GR, Benedict EQ (2020) Environmental aspects of densification of residential development in urbanized territories (case study: Kazan City). Casp J Environ Sci 18:519–524
36. Pellegrini P, Chen J (2020) Hypothesis of densification for a sustainable urbanization in a wealthy Chinese city. IOP Conf Ser Earth Environ Sci 588
37. Wang C, Li Y, Shen Y, Liu Y, Ru P, Wei Z, Xie D (2024) Addressing the influencing path of social noise exposure risk perception on noise mitigation behavior. J Environ Manag 353:120238. https://doi.org/10.1016/j.jenvman.2024.120238
38. Chen J, Pellegrini P, Xu Y, Ma G, Wang H, an Y, Shi Y, Feng X (2020) Evaluating residents' satisfaction before and after regeneration. the case of a high-density resettlement neighbourhood in Suzhou, China. Cogent Soc Sci 8:0–26. https://doi.org/10.1080/23311886.2022.2144137
39. Zhao X, Chen J, Li J, Wang H, Zhang X, Yu F (2025) Unraveling the renewal priority of urban heritage communities via macro-micro dimensional assessment-a case study of Nanjing City, China. Sustain Cities Soc 124:106317
40. Song H, Chen J, Li P (2025) Decoding the cultural heritage tourism landscape and visitor crowding behavior from the multidimensional embodied perspective: insights from Chinese classical gardens. Tour Manag 110:105180. https://doi.org/10.1016/j.tourman.2025.105180
41. Zhao X, Yu F, Zhang, X Chen J, Li P (2025) Assessing Urban Renewal Efficiency via Multi-Source Data and DID-Based Comparison between Historical Districts. npj Herit. Sci 13: 389, https://doi:10.1038/s40494-025-01964-9.
42. Pellegrini P, Chen J (2025) The Regeneration Path Not Taken Further: An Experiment in Urban Densification and State Entrepreneurialism in a Resettlement Neighborhood in Suzhou, China. Urban Stud. 00420980251365188, https://doi:10.1177/00420980251365188

Correction to: Densification as a Design Tool for Sustainable Urban Regeneration

Correction to:
J. Chen and P. Pellegrini, *Densification as a Design Tool for Sustainable Urban Regeneration*, Urban Sustainability, https://doi.org/10.1007/978-981-95-1653-7

In the original version of the book, the following belated corrections are to be incorporated: For all chapters, Associate Prof. Paola was designated as the corresponding author. In an earlier version of this paper, the acknowledgement was missing for Chapter 6. This has now been corrected. In an earlier version of this paper, the acknowledgement was missing for Chapter 8. This has now been corrected. In an earlier version of this paper, the acknowledgement was missing for Chapter 10. This has now been corrected. The original version of this chapter was revised: Author provided figure corrections for Chapter 10 have been incorporated. This has now been corrected. In an earlier version of this paper, the acknowledgement was missing for Chapter 11. This has now been corrected. The original version of this chapter was revised: Author provided figure corrections for Chapter 11 have been incorporated.

The updated version of this book can be found at
https://doi.org/10.1007/978-981-95-1653-7

J. Chen and P. Pellegrini, *Densification as a Design Tool for Sustainable Urban Regeneration*, Urban Sustainability, https://doi.org/10.1007/978-981-95-1653-7_13

FSC
www.fsc.org
MIX
Papier aus verantwortungsvollen Quellen
Paper from responsible sources
FSC® C105338

If you have any concerns about our products,
you can contact us on
ProductSafety@springernature.com

In case Publisher is established outside the EU,
the EU authorized representative is:
Springer Nature Customer Service Center GmbH
Europaplatz 3, 69115 Heidelberg, Germany

Printed by Libri Plureos GmbH
in Hamburg, Germany